T-Labs Series in Telecommunication Services

It is the aim of the Springer Series in Telecommunication Services to foster an interdisciplinary exchange of knowledge addressing all topics which are essential for developing high-quality and highly usable telecommunication services. This includes basic concepts of underlying technologies, distribution networks, architectures and platforms for service design, deployment and adaptation, as well as the users' perception of telecommunication services. By taking a vertical perspective over all these steps, we aim to provide the scientific bases for the development and continuous evaluation of innovative services which provide a better value for their users. In fact, the human-centric design of high-quality telecommunication services – the so called "quality engineering" – forms an essential topic of this series, as it will ultimately lead to better user experience and acceptance. The series is directed towards both scientists and practitioners from all related disciplines and industries.

** Indexing: books in this series are indexing in Scopus **

Ashutosh Singla

Assessment of Visual Quality and Simulator Sickness for Omnidirectional Videos

Springer

Ashutosh Singla
TU Ilmenau
Ilmenau, Germany

ISSN 2192-2810 ISSN 2192-2829 (electronic)
T-Labs Series in Telecommunication Services
ISBN 978-3-031-49987-6 ISBN 978-3-031-49988-3 (eBook)
https://doi.org/10.1007/978-3-031-49988-3

This Springer imprint is published by the registered company Springer Nature Switzerland AG
The registered company address is: Gewerbestrasse 11, 6330 Cham, Switzerland

Preface

This book presents extensive research on the quality of 360° video perceived by users with HMDs. Prior to this work, no such standard Recommendation existed for 360° videos. Hence, it is necessary to develop a set of guidelines for the systematic visual quality assessment of 360° videos. Firstly, conventional subjective test methods such as Absolute Category Rating (ACR) and Double Stimulus Impairment Scale (DSIS) were applied to evaluate video quality, alongside the Modified ACR (M-ACR) method newly proposed during the research. Building on the reliability and general applicability of the procedure across different tests, a methodological framework for 360° video quality assessment is presented in this book.

For 360° videos, an increased degree of immersion comes along with the problem of simulator sickness, as a further QoE constituent. Therefore, simulator sickness is analyzed in this book to investigate the impact of different influencing factors. The insights gained on simulator sickness related to 360° video contribute to a better understanding of this particular use case of VR and can help to improve comfort among users by suggesting improvements in the technical specifications of 360° video and HMD technology and thus improving QoE. Additionally, a simplified Simulator Sickness Questionnaire (SSQ) for the self-reporting of symptoms that are relevant for 360° videos is proposed, by comparing different versions of questionnaires with the state-of-the-art variants CSQ (Cybersickness Questionnaire) and VRSQ (Virtual Reality Symptom Questionnaire), alongside the legacy SSQ set of scales.

Furthermore, for omnidirectional video, exploration behavior and presence also represent key aspects of QoE. Hence, as an additional construct, initial research on presence was included for 360°-video evaluation. This way, a relationship between different considered constructs (media quality, simulator sickness, and presence) is developed. As an additional measure, subjects' viewing behavior was captured for all tests. As a result, media quality and simulator sickness are interpreted in light of subjects' viewing behavior.

This book is intended for students and researchers engaged in the evaluation of QoE for 360° videos. Moreover, it could prove valuable for individuals working on assessing simulator sickness in omnidirectional video contexts.

Ilmenau, Germany
February, 2023

Ashutosh Singla

Acknowledgments

I would like to offer my sincere gratitude to my supervisor, Prof. Dr.-Ing. Alexander Raake, who continuously supported me during the thesis. His valuable suggestions and guidance helped me in keeping myself motivated throughout my Ph.D. journey. He provided countless opportunities that were helpful for the development of my scientific career. I am very grateful to him for accepting me as his Ph.D. student and believing in me.

I would like to thank Prof. Kjell Brunnström and Prof. Cheng-Hsin Hsu. I am very grateful to them for providing me with their valuable suggestions and comments that helped me in improving the quality of my book.

During my Ph.D. journey, I met a lot of colleagues from the Audiovisual Technology Group who became friends. First of all, I would like to thank Steve Göring for his endless support in scientific discussions and for coming up with ideas. To Werner, I would like to thank you from the bottom of my heart for supporting me during the initial phase of my Ph.D. To Janto, thank you for improving my presentation skills and supporting me. I would like to extend my thanks to Rakesh Rao, Michal, Stephan, Frank, Dominik, Mauro, William, and Thomas Robotham. Thank you for the time that we spent together. Special thanks to Eckhardt Schön, it has been a privilege to share the office with you. Our discussions are full of fun and laughter. I thank you for that.

I would like to thank Bernd, Christiane, Detlef, Moni, Matthias, Thomas, and Torsten for their technical and administrative support.

I would like to thank my former colleagues and friends from the MMSPG lab: Martin, Philippe, Anne-Flore, Lin, Irene, Evgeniy, and Evangelos. It was amazing to know you guys. I would like to thank Prof. Touradj Ebrahimi for accepting me at the MMSPG lab and supporting me in pursuing doctoral studies.

I would like to thank my family: my father, mother, sisters, brothers-in-law, and especially Pratham, Fanty, Vamika, and Anisha, for their endless support.

Contents

Acronyms

ACR	Absolute Category Rating
ANOVA	Analysis of Variance
AVT	Audiovisual Technology Group
CI	Confidence Interval
CRF	Constant Rate Factor
CSQ	Cybersickness Questionnaire
D	Disorientation
DoF	Degrees of Freedom
DSIS	Double Stimulus Impairment Scale
EFA	Exploratory Factor Analysis
ERP	Equirectangular Projection
FHD	Full HD
FMS	Fast Motion Sickness
FOV	Field of View
FPS	Frames per second
HD	High Definition
HM	HEVC Test Model
HMD	Head Mounted Display
HRC	Hypothetical Reference Circuit
IPD	Interpupillary Distance
IPQ	igroup Presence Questionnaire
ITC-SOPI	ITC-Sense of Presence Inventory
ITQ	Immersive Tendencies Questionnaire
ITU	International Telecommunication Union
JEM	Joint Exploration Test Model
JSON	JavaScript Object Notation
JVET	Joint Video Exploration Team
KMO	Kaiser-Meyer-Olkin
M–ACR	Modified Absolute Category Rating
MASQ	Motion Sickness Assessment Questionnaire
MCI	Mean Confidence Interval

MI	Motion Interpolation
MISC	Misery Scale
MOS	Mean Opinion Score
MS	Motion Sickness
MUSHRA	Multiple Stimuli with Hidden Reference and Anchor
N	Nausea
O	Oculomotor
OS	Operating System
PC	Principal Component
PCA	Principal Component Analysis
PCC	Pearson's Correlation Coefficient
PQ	Presence Questionnaire
PVS	Processed Video Sequence
QHD	Quad High Definition
QoE	Quality of Experience
QP	Quantization Parameter
RCMP	Reshaped Cubemap Projection
RMSE	Root Mean Square Error
RTT	Network Round Trip Time
SAMPVIQ	Subjective Assessment of Multimedia Panoramic Video Quality
SAMVIQ	Subjective Assessment of Multimedia Video Quality
SDK	Software Development Kit
Short PQ	Short Presence Questionnaire
Short SSQ	Short Simulator Sickness Questionnaire
SI	Spatial Index
SiSiMo	Simulator Sickness Model
SRCs	Source Sequences
SSCQS	Single Stimulus Continuous Quality Scale
SSQ	Simulator Sickness Questionnaire
STD	Standard Deviation
TI	Temporal Index
TS	TotalScore
TSP	Truncated Square Pyramid
UHD	Ultra High Definition
VBV	Video Buffering Verifier
VE	Virtual Environment
VIMS	Visually Induced Motion Sickness
VR	Virtual Reality
VRSQ	Virtual Reality Sickness Questionnaire
VRSQ	Virtual Reality Symptom Questionnaire
VVC	Versatile Video Coding

Chapter 1
Introduction

In the mid-1930s, Stanley G. Weinbaum, an American science fiction writer, envisioned in his story "Pygmalion's Spectacles" a pair of goggles capable of giving a person sight, sound, taste, smell, and touch (Weinbaum 2016). With the advent of modern Head-Mounted Displays (HMDs), such as HTC Vive Pro, Oculus Rift, Pimax, etc., researchers have been successful in making his vision a reality.

HMDs enable to present audiovisual information to the users' ears and eyes that can be explored interactively by orienting within the displayed scenes via head and eye movements. In applications such as games with computer-generated scenes, also full-body translational movements are enabled for scene exploration, or interaction with objects in the scene, for example, via hand controllers. Human interaction with a physically real or virtual-reality-generated scene involves a close coupling of sensorimotor control with audiovisual sensory information. In the case of HMD-based virtual reality, all elements of the scene-generation process decide on the virtual reality experience provided to the user. Different influencing factors determine the experience of users of such a VR system.

One prominent use case for current VR technology involving HMDs is 360° video. Such 360° video can enable a more immersive experience than traditional 2D-screen-based viewing. With a 360° scene, screen-typical eye movements across a rectilinear viewport are complemented by natural head movements to explore the scene. For traditional 2D video, user-centric technology assessment typically focuses on video quality or audiovisual-media-quality evaluation. Respective well-established test methods with human subjects or computational quality prediction approaches cannot be directly applied to the domain of 360° video quality. As known from 3D video QoE evaluation, novel assessment approaches are required (Sanchez 2016; Dehkordi 2015). Similarly, for 360° video, the increased degree of immersion may come along with the problem of simulator sickness. This effect may occur in case the motion present in a given video does not correspond to the motion information provided by sensorimotor cues involved in head and eye movements and by the vestibular system.

A. Singla, *Assessment of Visual Quality and Simulator Sickness for Omnidirectional Videos*, T-Labs Series in Telecommunication Services,
https://doi.org/10.1007/978-3-031-49988-3_1

In the proposed work, the virtual reality experience is broken down into three inter-related components: (i) visual quality, (ii) presence, and (iii) comfort, which comprises, as the most prominent aspect, simulator sickness (Kennedy et al. 1993) related to the viewing experience. It can be assumed that these three constructs (i–iii) are inter-related. Investigating all three components for a variety of different VR-related use cases can be considered infeasible for an individual book. Hence, the proposed book will primarily focus on the usage of head-mounted display-based VR technology in the context of 360° video viewing sessions.

To reliably assess the visual quality of 360° videos, standardized test methods are needed. However, prior to this work, no such standard recommendation existed for 360° videos. Hence, it is necessary to develop a set of guidelines for the systematic visual quality assessment of 360° videos. In addition, while watching 360° videos or interacting with a VR environment wearing an HMD, users may experience simulator sickness. At present, practically no solution exists that can effectively and efficiently eradicate the symptoms of simulator sickness (or cybersickness) from virtual environments. Therefore, in the absence of such a solution, it has become indispensable to further reduce the amount of simulator sickness, by making improvements in the technical specifications of the used technology and video content. Furthermore, several test methods exist to self-report the symptoms of simulator sickness. One such method for self-reporting the symptoms of simulator sickness is the SSQ (Kennedy et al. 1993). The SSQ was designed mainly for measuring simulator sickness experienced by Navy pilots for flight simulator studies. There are differences between simulator and VR systems, and it may be possible that some of the symptoms that are relevant for simulator-system studies may not be relevant for VR, and in particular for 360° video. It could be possible that users may feel less prone to simulator sickness in 360° videos than in simulator studies. Therefore, it is reasonable to investigate if it is efficient to use the SSQ in its current form for 360° videos.

1.1 Contributions

The main contributions of this book are summarized below:

1.1.1 Visual Quality Assessment of 360° Videos

Extensive research on the quality of 360° video perceived by users with HMDs is presented in this book. Firstly, conventional subjective test methods such as ACR ITU-R BT.500-13 (2012) and DSIS ITU-T Rec. P.910 (2008) were applied to evaluate video quality, alongside the M-ACR Singla et al. (2017) method newly proposed during the book research. Building on the reliability and general applicability of the procedures across different tests, a methodological framework for 360° video quality assessment is presented.

Moreover, different sets of "bitrate-ladder[1]" type curves for video quality over bitrate and resolution were derived for different subjective tests, using different test methods to derive the optimal bitrate and resolution for the current generation of HMDs. This will help in saving bandwidth.

Recommendations/guidelines were derived for conducting tests for the systematic evaluation of media quality of 360° videos. Prior to this work, no such guidelines existed for the evaluation of 360° video quality. ITU-T P.919 (2020) was the first recommendation on the subjective test methods for 360° video—with contributions from the author.[2]

1.1.2 Simulator Sickness

One side effect of 360°/VR videos is that it may lead to undesirable effects on the user's well-being, referred to as cybersickness or simulator sickness. Besides the individual implications on viewing comfort, cybersickness may reduce the viability of VR technology and its applications. Therefore, it is important to know how cybersickness can be assessed, predicted, and minimized.

In this book, the influence of different factors such as video resolution, bitrate, session duration, and latency on cybersickness has been investigated. It was shown that a more pleasant experience in HMDs can be achieved by making improvements in technical factors such as resolution and bitrate. It has been shown that in a higher resolution video, users are less prone to simulator sickness. Additionally, users are more comfortable watching a higher bitrate video than in a lower one.

Furthermore, a conceptual framework called Simulator Sickness Model (*SiSiMo*) to develop models to predict simulator sickness scores has been proposed in the course of this chapter. This framework takes into account various transitions during a 360° video viewing session such as viewing, rating, pausing, and adaptation between these steps. As a first step towards developing such a model that considers these transitions, an important contribution of this book is a first model instance with regard to the temporal evaluation of simulator sickness. It was demonstrated that the temporal evolution of discomfort can be estimated using linear regression over viewing-time index for 360° videos.

[1] https://netflixtechblog.com/per-title-encode-optimization-7e99442b62a2 (Accessed on 2017/03/15).

[2] The author made contributions in Sections 3–11.

1.1.3 Assessment of Simulator Sickness Questionnaire

The SSQ was designed mainly for measuring simulator sickness experienced by Navy pilots for flight simulator studies. Since the SSQ with its 16 questions was not designed for 360° video related studies, it was hypothesized during the course of this book that it may be simplified to enable more efficient testing for 360° video. Hence, the SSQ was evaluated with the aim to reduce the number of questions that can potentially be asked to subjects for self-reporting of simulator sickness.

Based on the conducted analysis, it was shown that nine symptom-related attributes appear to be sufficient to represent the three dimensions of SSQ, as some attributes included in the original SSQ are not relevant for 360° video studies. These nine symptoms can be expected to capture all the symptoms relevant for 360° video, just as the SSQ does, and hence can be used as an alternative to the SSQ for studies related to 360° video.

1.1.4 Holistic Evaluation of QoE for 360° Video

Due to the immersive nature of 360° videos, QoE for these videos comprises additional components such as simulator sickness and presence as compared to classical 2D videos. Therefore, besides media quality, simulator sickness and presence need to be assessed for a holistic understanding of QoE in the case of omnidirectional videos. Here, different test protocols were compared to assess integral quality, simulator sickness, and presence for omnidirectional videos. The book analyzes the usage of both traditional questionnaires and single-scale approach for presence and simulator sickness, and from the order of asking judgments from test subjects using both within- and between-subject design. Based on the applied mixed within- and between-subject design, it is possible to analyze relations between scale usages as well as order effects and statistical power of the tests. Further, the interrelation between the ratings for quality, presence, and simulator sickness is investigated.

1.1.5 Reproducible Research

All of the data related to video quality, presence, simulator sickness, exploration behavior, and video sequences of the different tests conducted during the course of this book are made publicly available.[3] This approach ensures that all the results published in the book are reproducible. Furthermore, to summarize each step of

[3] https://avtshare01.rz.tu-ilmenau.de/asingla/PhDData/.

different evaluation experiments, the research work was published in peer-reviewed conferences such as QoMEX, ACM MM, HVEI, IEEE VR, MMSP, and MMSys, and standardization meetings such as JVET, and ITU-T.

1.2 Organization of Book

Chapter 1 (Introduction) provides the main motivation, contribution, and organization of the book.

Chapter 2 (State of the Art) describes the different terminologies associated with 360° video, simulator sickness, presence, and behavior. Furthermore, an overview of the state of the art for 360° datasets, evaluation of 360° video quality, of simulator sickness and of presence, and of exploration behavior is provided.

Chapter 3 (Technical Setup) describes the pipeline of the VR testing system and presents the details of the source sequences used in the subjective tests conducted as part of this book. Further, this chapter provides a brief overview of the technical environment such as HMDs, the PC, and the framework for recording behavioral data.

Chapter 4 (Comparison of Different Subjective Test Methods) presents the results of two different experiments conducted in two different contexts in a lab setting, where the perceived quality of 360° videos was evaluated and compared using different test methods for different resolutions and bitrates.

Chapter 5 (Visual Quality Assessment of 360° ***videos****)* investigates the impact of HMDs, encoding settings (resolution and bitrate), voting methods, and different delay values on tile-based streaming in relation to the perceived video quality. Furthermore, based on the state of the art and the results presented in this chapter, recommendations were provided for the improvement of the perceived visual quality.

Chapter 6 (Simulator Sickness Assessment for 360° ***Videos****)* focuses on evaluating the impact of different influencing factors such as video resolution, bitrate, session duration, and latency on cybersickness. Furthermore, a conceptual framework called "SiSiMo" is presented, which can be used to develop a model to estimate/predict simulator sickness over longer viewing sessions. A first model instance is described based on our test data.

Chapter 7 (Performance Analysis of SSQ) investigates the efficiency of the SSQ in its current form for 360° videos, deriving a reduced set of scales to enable a more dedicated and time-efficient assessment. The simplified version of the SSQ proposed in this chapter focuses on the symptoms that are relevant for 360° video related studies, with the aim to reduce test time and ultimately enable more expansive testing.

Chapter 8 (Evaluation of Different Constructs—Media Quality, Simulator Sickness, and Presence) investigates video quality, simulator sickness, and presence for 360° video. Here, in particular, the suitability of short, single-scale versions of the simulator sickness and presence questionnaires in comparison to the long versions of SSQ (Kennedy et al. 1993) and PQ (Witmer and Singer 1998) is studied. Further, the interrelation between the ratings for quality, presence, and simulator sickness is investigated.

Chapter 9 (Conclusion and Future Work) summarizes the contributions of the book and provides an outlook on possible future work.

References

A.B. Dehkordi, *3D Video Quality Assessment*. Doctoral Dissertation, The University of British Columbia, 2015

R.S. Kennedy et al., Simulator sickness questionnaire: An enhanced method for quantifying simulator sickness. Int. J. Aviat. Psychol. **3**(3), 203–220 (1993)

ITU-T Rec. P.910, *Recommendation ITU-T P.910: Subjective Video Quality Assessment Methods for Multimedia Applications* (2008)

ITU-R BT.500-13, *Recommendation ITU-R BT.500-13: Methodology for the Subjective Assessment of the Quality of Television Pictures* (2012)

ITU-T P.919, *Subjective Test Methodologies for 360° Video on Head-Mounted Displays*. Recommendation P.919 (2020)

J.G. Sanchez, *Analysis of Quality of Experience in 3D Video Systems*. Doctoral Dissertation, Polytechnic University of Madrid, 2016

A. Singla et al., Comparison of subjective quality evaluation for HEVC encoded omnidirectional videos at different bit-rates for UHD and FHD resolution, in *Proceedings of the on Thematic Workshops of ACM Multimedia*, Mountain View, California, USA (2017). ISBN: 978-1-4503-5416-5. https://doi.org/10.1145/3126686.3126768. http://doi.acm.org/10.1145/3126686.3126768

S.G. Weinbaum, *Pygmalion's Spectacles* (Simon and Schuster, New York, 2016)

B.G. Witmer, M.J. Singer, Measuring presence in virtual environments: a presence questionnaire. Presence **7**(3), 225–240 (1998)

Chapter 2
State of the Art

Measuring and validating QoE of omnidirectional videos using different HMDs has become increasingly important. The resulting user experience plays an important role in making technology successful and acceptable for a wide user group. The respective "QoE" is defined as the "...degree of delight or annoyance of the user of an application or service..." Le Callet et al. (2012) ITU–T Rec. P.10/G.100 (2017). One component of QoE is the perceptual media quality of the viewing session, resulting from different technical characteristics of the end-to-end chain comprising the recording, post-processing, representation, coding, retrieval, transmission, playback, and display of 360° videos. Furthermore, for 360° video, exploration behavior, presence, and simulator sickness also represent key aspects of QoE. Therefore, besides media quality, these three further constructs need to be assessed for a holistic understanding of QoE in the case of omnidirectional videos.

2.1 Preliminary Concepts

In this section, a brief overview will be provided related to the terminologies associated with 360° video, simulator sickness, and presence.

360° video (also known as omnidirectional video) is captured from a real-life scene using a 360° camera and can provide movement to a user in only 3 DoF (3 rotational). In contrast to 360° video, VR can provide movement to a user in 6 DoF (3 directional and 3 rotational).

Motion Sickness (MS) and Visually Induced Motion Sickness (VIMS) have similar symptoms but are caused due to different reasons (Kolasinski 1995). MS can occur because of actual motion, and VIMS is caused due to visual stimulation. Simulator sickness and cybersickness are similar to motion sickness but caused

A. Singla, *Assessment of Visual Quality and Simulator Sickness for Omnidirectional Videos*, T-Labs Series in Telecommunication Services,
https://doi.org/10.1007/978-3-031-49988-3_2

by visual stimuli and can therefore also be referred to as VIMS[1] Johnson (2005). Cybersickness and simulator sickness are similar, but they have different symptom profiles. Cybersickness is caused when people are exposed to a Virtual Environment (VE), and simulator sickness occurs due to effects associated with a simulator (Hale and Stanney 2014; Kolasinski 1995; LaViola 2000; Kennedy et al. 2003).

Presence is defined as the experience of being in one place when a person is physically situated in another (Witmer and Singer 1998; Slater 2003). Immersion is a psychological state characterized by perceiving oneself to be enveloped by, included in, and interacting with an environment that provides a continuous stream of stimuli and experiences (Witmer and Singer 1998). Involvement is a psychological state experienced as a consequence of focusing one's energy and attention on a coherent set of stimuli or meaningfully related activities and events (Witmer and Singer 1998). For experiencing presence, both involvement and immersion are necessary (Slater 2003).

While watching 360° videos with an HMD, users can explore the content using head and eye movements. For the present book, only head tracking will be focused on. The head rotation data are captured in three co-ordinates referred to as yaw, pitch, and roll.

2.2 360° Datasets

In recent years, several 360° video datasets have been published (Abbas and Adsumilli 2016; Asbun et al. 2016; Schwarz et al. 2016; Sun and Guo 2016; Duan et al. 2017; Lo et al. 2017a; Li et al. 2017; Nasrabadi et al. 2019; Mahmoudpour and Schelkens 2019). In Abbas and Adsumilli (2016), Asbun et al. (2016), Schwarz et al. (2016), Sun and Guo (2016), the 360° dataset is available in an uncompressed format or raw videos, without audio. The main limitation of their sequences is the duration (10 s), and the scenes are focused on the aspect of video quality. In turn, the 10 s videos are too short to meaningfully investigate the effect of simulator sickness and presence on the users as it may take some time to develop the symptoms. Furthermore, 360° video sequences can also be downloaded from online platforms such as YouTube, Arte, or ImmersiaTV;[2] however, they are not available in uncompressed or very high-quality versions. Duan et al. (2017) published a 360° dataset recorded with the stationary Insta360 4K that is publicly available. Their dataset contains 10 video sequences with audio in 4K resolution with a duration of 15 s. Lo et al. provided a dataset taken from YouTube (Lo et al. 2017a). Their dataset contains 10 videos in 4K resolution at 30 frames per second (fps). Along with this dataset, the authors also provided saliency maps, motion maps, and positional and

[1] Please note that the word "sickness" does not mean here as a disease, but it is referred here as a response of an individual when exposed to real motion or to stimuli (Johnson 2005).

[2] http://www.immersiatv.eu/project-outcomes/datasets/.

rotational data of 50 users. Li et al. (2017) created the dataset from the websites such as YouTube, Vrideo, and Facebook. They selected 73 videos that ranged from 29 to 668 s. Besides videos, they provided valence, arousal, and the rotational head movements of the 95 participants.

Furthermore, Nasrabadi et al. (2019) published a 360° dataset. Their dataset contains sequences from YouTube, and six sequences are recorded with Samsung Gear360 in 4K resolution without audio with a duration of 60 s. Mahmoudpour and Schelkens (2019) provided 15 360° videos with a GoPro Fusion 360° camera with a resolution of 3840×1920 at 30 fps with a duration of 12 s.

Most of the datasets described above have selected the videos from YouTube, and only a few of the datasets were captured by themselves using 360° camera. A new and high-quality 360° video sequences are produced in this book (see Sect. 3.3). The proposed sequences have a duration of at least 30 seconds and cover a wide range of spatial properties and exploration behavior.

2.3 360° Video Quality

For 360° video, user-centric technology assessment typically focuses on video quality or audiovisual-media-quality evaluation. Relevant system influencing factors for the QoE of omnidirectional videos are, for example, compression algorithms, projection schemes, display devices along with their rendering technologies, resolution, bitrates, etc. Hence, assessing the impact of these factors on the QoE of omnidirectional videos with appropriate methods is essential. In this section, mainly subjective video quality will be addressed. Several studies have been reported in the state of the art dedicated to the assessment of subjective video quality (Tran et al. 2017; Duan et al. 2017; Zhang et al. 2017, 2018; Li et al. 2018; Lopes et al. 2018; Hanhart et al. 2018; Mahmoudpour and Schelkens 2019; Covaci et al. 2019; Hofmeyer et al. 2019; van Kasteren et al. 2020; Gutierrez et al. 2022). Table 2.1 summarizes several studies conducted for the assessment of the visual quality of 360° videos.

In Tran et al. (2017), Tran et al. investigated the impact of different resolutions and bitrates on the visual quality of 360° videos with Samsung Gear VR using the ACR method, using the classical 5-point scale. Experimental results showed that 4K and Quad High Definition (QHD) resolutions can provide better-perceived quality than full High Definition (FHD) and High Definition (HD). Furthermore, they found that Samsung Gear VR (S6 phone) provides a better-perceived quality than Google Cardboard (S5 phone) irrespective of resolution and video sequence. In Duan et al. (2017), Duan et al. investigated the impact of frame rate, resolution, and bitrate using the ACR test method, establishing a dataset that contains 10 raw videos in 4096×2048 shot using an Insta360. Experimental results showed that higher resolution (4096×2048) and higher frame rate (30 fps) provide better visual quality than lower resolutions (2048×1024 and 1024×512) and frame rates (5 and 15 fps). Furthermore, the perceived visual quality at 70,000 KBit/s is slightly higher than at 10,000 KBit/s.

Table 2.1 Summary of several tests for 360° video quality assessment. Some of the entries in this table are taken from Table 1 Singla et al. (2020)

Name	Method	Resolution	Frame rate	Encoder	HMD name	No. of subjects	Duration of video
Tran et al. (2017)	ACR	4K, QHD,	30	H.264/AVC	Gear VR (S6)	36	30 s
		FHD, HD			Google Cardboard (S5)		
Duan et al. (2017)	ACR	4K, 2K, 1K	30, 15, 5	–	HTC Vive	13	15 s
Xu et al. (2017)	SSCQS (ACR)	4K	25	HEVC	HTC Vive	48	12 s
Zhang et al. (2017)	SSCQS, SAMVIQ	4K	30	H.264, H.265	HTC Vive	23	10 s
	SAMPVIQ			VP9			
Li et al. (2018)	SSCQS	4K–8K	24–30	HEVC	HTC Vive	221	10–23 s
Lopes et al. (2018)	ACR &	R1, FHD,	7.5, 10, 15	HEVC	Oculus Rift	37	10 s
	ACR-HR	4K, 8K	30, 60				
Zhang et al. (2018)	ACR-HR	3600×1800	30, 60	HEVC	HTC Vive	30	10 s
Hanhart et al. (2018)	DSIS	8K	30	HM and JEM	2D Screen	8	10 s
Mahmoudpour and Schelkens (2019)	ACR	4K	30	HEVC	Oculus Rift	20	12 s
Covaci et al. (2019)	ACR	HD, FHD,	25–60	H.264	Samsung Gear VR	48	60 s
		2.5K, 4K					
Hofmeyer et al. (2019)	ACR, PC	4K	25, 30, 50,	H.265	HTC Vive	12	20 s
			60, 90		Pro		
Yao et al. (2019)	ACR	4K	30	–	Rift DK2	60	30
Anwar et al. (2020)	ACR	4K, 2.5K,	24–30	H.264	HTC Vive	29	60 s
		FHD, HD			Google Cardboard (S A7)		
Cortés et al. (2020)	DCR	4K	–	–	Lenovo Mirage Solo	20	60 s
van Kasteren et al. (2020)	ACR	4K	30	H.264	HTC Vive	33	–
Gutierrez et al. (2022)	ACR, DCR	4K	25, 30	H.265	Vive, Rift	306	10, 20, 30
					Vive Pro, Gear VR		

In Zhang et al. (2017), Zhang et al. compared the performance of three subjective test methods: SSCQS, subjective assessment of multimedia video quality (SAMVIQ), and subjective assessment of multimedia panoramic video quality (SAMPVIQ). Experimental results show that SAMPVIQ has a better Pearson's correlation coefficient (PCC) with the video quality metric (VQM) (Pinson and Wolf 2004) used for comparison. However, the authors have not validated which test method is statistically more reliable. Also, it is not clear if these three test methods are compared under the same contextual conditions.

In Li et al. (2018), Li et al. compared the three different projection methods equirectangular projection (ERP), reshaped cubemap projection (RCMP), and truncated square pyramid projection (TSP), using a continuous quality scale with HTC Vive. Results suggest that TSP provides better subjective video quality. Further, it is seen that bitrate has more impact on subjective quality as compared to projection schemes. In Lopes et al. (2018), Lopes et al. evaluated the impact of different resolutions (R1 (960×480), FHD, 4K, 8K) and frame rates (7.5, 10, 15, 30, and 60) on the visual quality of several omnidirectional videos using ACR and ACR-HR (ACR with Hidden-reference Removal) with an Oculus Rift. Experimental results indicate that 4K and 8K resolution provided similar perceived quality at 30 fps. Also, results showed that there is a saturation in visual quality when temporal resolution increased from 30 to 60 fps at 4K resolution. It is noted that the lack of differences between 4K and 8K resolution may be a result of either the source contents and their respective quality or of the limited resolution and resulting screen-door effect for the Oculus Rift used (Sitter et al. 2017).

In Zhang et al. (2018), Zhang et al. proposed a protocol for the subjective assessment of omnidirectional videos for coding applications. Considering the projection and resolution of the applied HMD, video sequences should be re-sampled to the optimal resolution 3600×1800 before coding for HTC Vive to alleviate the interference of HMD sampling. In Hanhart et al. (2018), Hanhart et al. evaluated the visual quality for four 360° video sequences in 8K resolution on a 2D screen for the HM (HEVC Test Model) and JEM (Joint Exploration Test Model) reference encoders using the DSIS method. For each sequence, two dynamic viewports were extracted. Their results showed that JEM performs better at the lower and intermediate bitrates, irrespective of the sequence, whereas the performance of HM and JEM is similar at the highest bitrate.

In Mahmoudpour and Schelkens (2019), Mahmoudpour et al. studied the impact of judder and compression artifacts on the subjective visual quality of omnidirectional videos in Oculus Rift using a Single-Stimulus Continuous Quality Evaluation (SSCQE) (ITU-R BT.500-13 2012) method. A dataset[3] was created that contains 12 omnidirectional videos shot with a GoPro Fusion 360° camera and three computer-generated VR videos. The test results show that at higher quality levels, the quality drop due to judder is more significant than that due to compression, at least for the considered parameter ranges. In Covaci et al. (2019), Covaci et al. investigated

[3] http://data.etrovub.be/qualitydb/judder-vqa.

the impact of mulsemedia (multisensory media) on 360° videos. Experimental results showed that users perceived better quality for 360° mulsemedia than 360° multimedia irrespective of encoding quality. In Hofmeyer et al. (2019), Hofmeyer et al. investigated several combinations of 360° players and frame rates. They found that 90 fps provides significantly better quality than other frame rates (25, 30, 50, 60). Further, results showed that *Whirligig* and *Virtual Desktop* both offer a smooth playback for 360° videos without flickering at 30 and 90 fps. They also investigated the impact of various motion interpolation (MI) algorithms on QoE, when interpolating (temporally upscaling) sequences from 30 to 90 fps. The original frame rate was 30 fps. It was concluded that MI, in general, is a suitable tool for increasing the QoE of 360° videos. While FFmpeg MCI should be preferred over butterflow for videos containing fast and sudden movements, for slow- and medium-motion videos, butterflow is more suitable.

The impact of user's interest, familiarity, quantization parameter (QP), resolution, and rendering device on perceptual quality was investigated by Anwar et al. (2020). Their experimental results showed that the user's interest, familiarity, QP, resolution, and rendering device have an impact on the perceptual quality. The impact of network delay was investigated by Cortés et al. (2020), and their results showed that an increase in delay decreases the quality scores irrespective of camera systems. The impact of quality degradation and freezing was investigated for three 360° video sequences on QoE in van Kasteren et al. (2020). Their experimental results showed that users are more sensitive to freezing events as compared to quality degradation using the Constant Rate Factor (CRF). Users provided lower quality ratings in the case of freezing events, even at the highest CRF. A user study was conducted with different projection schemes (ERP, Adjusted Equal-area Projection (AEP), and Equiangular Cubemap Projection (ECP)) and two different QPs to develop QoE models for watching videos in an HMD. Their model has a moderate performance with a Pearson coefficient of 0.71.

In Gutierrez et al. (2022), several subjective tests were conducted by different research labs (including ours) using different test methods (ACR and DCR), different HMDs (Oculus Rift, HTC Vive, Vive Pro, and Gear VR), with different sequence durations (10 s, 20 s, and 30 s). Their results showed that there is a significant effect of test methodology for ACR with respect to DCR. However, from the results, it is not clear which of the test method is more reliable. Furthermore, the influence of sequence duration was investigated too. The ACR 10 s vs. 20 s, DCR 10 s vs. 20 s, ACR 10 s vs. 30 s, and DCR 20 s vs. 30 s are statistically significant. The influence of HMD was also addressed. Their analysis showed no significant difference between Gear VR and HTC Vive. However, significant differences were found comparing HTC Vive and HTC Vive Pro. The impact of audio was explored here. It was found that sequences without audio received higher ratings as compared to the sequences with audio.[4]

[4] No spatial audio was used.

Most of the studies reported above have used the ACR test method to evaluate the quality of 360° videos. The advantages of using the ACR method are that the Processed Video Sequence (PVS) is displayed only once, and it, therefore, removes the bias by removing the reference video. Furthermore, it allows the subjects to rate the quality independently. The state of the art lacks in finding a suitable evaluation methodology. The author's work presented in Chap. 4 aims at finding the suitable evaluation methodology for the assessment of the visual quality of 360° video.

2.4 Tile-Based Streaming

In viewport-adaptive streaming of omnidirectional video, a possible solution may be that only the field of view is streamed in high quality and the rest in a lower quality (Graf et al. 2017; Corbillon et al. 2017a; Gaddam et al. 2016; Petrangeli et al. 2017; Fan et al. 2019). This has significant benefits over streaming the entire 360 sphere (Lo et al. 2017b). A number of such systems leverage the support of tiles in modern coding systems, such as HEVC (Sullivan et al. 2012; Misra et al. 2013).

The time it takes between the users' head rotation and the display of high-quality tiles in the viewport is referred to as Motion-to-High-Resolution Latency (referred to as "delay" in the remainder of this book) (van Brandenburg et al. 2017). Note that this delay is determined by many factors, e.g., decoding delay, buffering, etc., see van Brandenburg et al. (2017). QoE testing is important, as tile-based viewport-adaptive streaming technologies are replacing classical approaches, because of significant bandwidth savings and an increase in viewing quality (Lo et al. 2017b). In recent years, several algorithms have been published proposing different solutions for tile-based 360° video streaming (Gaddam et al. 2016; Graf et al. 2017; Corbillon et al. 2017a; van Brandenburg et al. 2017; Monnier et al. 2017; Yen et al. 2019).

For 360° videos, efficient bandwidth utilization through adaptive streaming technology is a key factor to increase the user's perceived quality, being an important aspect to consider for network providers (van Brandenburg et al. 2017; Monnier et al. 2017). Different techniques are employed to save bandwidth, including optimization of video codecs (Graf et al. 2017), different tiling schemes (Zare et al. 2016), newer transmission protocols such as HTTP/2 (Petrangeli et al. 2017; Yen et al. 2019), prediction of viewer motion (Bao et al. 2016), viewport quality adaption (Hosseini and Swaminathan 2016), and more.

Along with effects due to reduced bandwidth, tile-based streaming of 360° videos introduces distortion during playout, e.g., tile switching delay (Ghosh et al. 2017), Motion-to-High-Resolution Latency. It is therefore important to be able to evaluate the Quality of Experience (QoE) of 360° streaming. Such QoE analysis can be done using objective metrics (Gaddam et al. 2016; Corbillon et al. 2017a), such as PSNR, SSIM variants, or others; however, such metrics are not always capable of measuring a user's true perception of quality. For example, the previously mentioned metrics are optimized for still-image quality, where temporal influences such as stalling (temporary interruption of media playback) or quality switches are not considered.

In Alface et al. (2012), the authors evaluated streaming performance at different bandwidth capacities with different delays. They measured the viewport PSNR as a performance measure and showed that improvement in video quality is possible, even with bandwidth constraints and interactive delays. However, their results were not validated by subjective tests. In Schatz et al. (2017), the authors studied the impact of stalling on the QoE of Omnidirectional video streaming and compared it with traditional 2D video. Results indicate that stalling has an impact on QoE. The authors found unexpected results in their study that may be an indication of an unreliable experimental setup; the authors do not explain the nature of the unexpected results. Curcio et al. (2017) presented subjective test evaluation results for viewport-dependent 360° video streaming. They compared two scenarios: (1) non-visible tiles were streamed at a lower resolution, and (2) non-visible tiles were streamed at full resolution, but at a lower signal-to-noise ratio (SNR). In their study, streaming the non-visible tiles at full resolution, but with lower SNR, provided better subjective video quality. In Upenik et al. (2016), Upenik et al. proposed a testbed for subjective quality evaluation of omnidirectional images in two different projection formats (Equirectangular and Cubemap) for different test methods. Their testbed is suitable for conducting the subjective test for omnidirectional images only.

The solutions presented above show that it is possible to stream omnidirectional videos at higher resolution (6K and above), without significantly increasing bandwidth and decoding requirements. However, there is a lack of experiments with a subjective quality evaluation of tile-based streaming QoE, particularly related to the impact of different Motion-to-High-Resolution Latencies on the perceived video quality, considering different scenarios. Moreover, none of the reviewed studies addresses the aspect of simulator sickness.

2.5 Simulator Sickness

While viewing 360° videos on an HMD, users may experience symptoms of simulator sickness such as nausea, dizziness, vertigo, sweating, etc. (Kennedy et al. 1993). Simulator sickness may occur during and after exposure to a virtual environment (VE) (Dużmańska et al. 2018). Hence, for a holistic picture of omnidirectional video QoE, simulator sickness should be monitored during respective tests.

There are different theories that explain why any one user may feel prone to the symptoms of simulator sickness. One of the most popular ones is the sensory conflict theory (or sensory mismatch theory) (Reason and Brand 1975; Duh et al. 2004). This theory states that the motion information that we receive from our visual senses (eyes) and vestibular system (inner ear) are integrated, and if these do not match, sickness may occur. For example, when a person is watching a video in an HMD, the eyes may be presented with content interpreted as motion, but the vestibular system is not, which leads to cybersickness. There are several other theories such as Treisman's hypothesis or the "Poison theory" (Treisman 1977) and Postural Instability Theory (Riccio and Stoffregen 1991) proposed in the literature to explain the phenomenon of simulator or cybersickness.

2.5.1 Influencing Factors

This section will provide a brief overview of how the symptoms of simulator sickness could be reduced by making improvements in the technical specifications of the used technology and video. The symptoms of simulator sickness could possibly be reduced if not eliminated completely by making advancements in tracking, resolution of the device, rendering solutions, low latency, and higher refresh rate, or by having users adapt to VE.

First, the effect of FOV was studied in relation to simulator sickness by Stern et al. (1990), Arthur (2000), and Chang et al. (2020). They showed that a wider FOV leads to an increase in the susceptibility of simulator sickness as it will increase the likelihood of flicker to be perceived at the peripheral vision (Maxwell 1992). One reason for this is that the human visual system at its periphery is more sensitive to flicker (Boff and Lincoln 1988). In order to avoid flicker, a higher refresh rate is required for a wider FOV (Maxwell 1992). In the case of slower refresh rates, users perceive more flicker that can cause the symptoms of nausea and eye strain or fatigue (Pausch et al. 1992). With commercially available HMDs, it will be difficult to investigate the impact of only refresh rate on the simulator sickness ratings because the other parameter such as display resolution, and FOV need to be constant. There are no two HMDs in the market that have the same technical specification except refresh rate. Therefore, to the best of the authors' knowledge, the exact impact of refresh rate on simulator sickness scores is unknown.

Duh et al. (2001) investigated the impact of six different FOVs (30°, 60°, 90°, 120°, 150°, and 180°) on the postural stability by projecting computer-generated scenes on the dome, which has a 180° × 180° FOV. Their experimental results proved that wider FOV leads to a greater imbalance in posture. DiZio and Lackner (1997) observed that subjects reported more symptoms of simulator sickness when using a wide FOV HMD. It is clear from the studies reviewed above that wider FOV results in a higher level of simulator sickness. Conversely, it provides more immersion as well (Duh et al. 2001). Rebenitsch and Owen (2021) and Chang et al. (2020) also recommended shrinking FOV for persons highly susceptible to sickness.

Lag (motion-to-photon latency) is defined as the time taken between user moves their head and this motion is being updated on the HMD. It is another important system factor that can influence simulator sickness (LaViola 2000; Pausch et al. 1992). Mania et al. (2004) evaluated the effect of lag on human perception and showed that users are able to perceive the lag beyond 15 ms. However, they did not evaluate the effect on the symptoms of simulator sickness. Brunnström et al. (2020) studied the impact of latency on video quality and comfort scores. Their experimental results showed that as the delay increases, the MOS (Mean Opinion Score) decreases, except at 55 ms. Further, comfort scores were reduced too with the increase in the delay. Wilson studied the effect of varying and constant latency over time in Wilson (2016). His results showed that as delay increases, simulator sickness also increases significantly in both the delay conditions. Further, they found that in the case of constant latency, users feel less prone to simulator sickness as compared to the varying latency. Chang et al. (2020) concluded that if the delay is constant while watching a video in an HMD, users can adapt to that delay. However, if there is a variable delay, then simulator sickness will be more. Their conclusions are matching with the ones in Wilson (2016). Therefore, it can be concluded that with the increase in delay, simulator sickness scores also increase, and that sickness scores are higher in the time-varying delay as compared to the constant delay.

Display Resolution: Guna et al. (2019) compared simulator sickness scores for different HMDs such as Oculus Rift DK 1 (640×800 per eye), Oculus Rift DK 2 (960×1080 per eye), and Oculus Rift CV 1 (1080×1200). Experimental results showed that users feel less prone to simulator sickness as the display resolution of the HMD increases. However, when comparing the technical specifications of all three versions of Oculus Rift, they are different in terms of display resolution, refresh rate, and FOV. Therefore, it cannot be deduced if the reduction in simulator sickness scores is due to the display resolution or refresh rate or FOV, or all.

Philipp (2020) investigated the impact of spatial and stereo audio. Their results proved that users are less prone to simulator sickness in spatial audio than in stereoscopic audio. The impact of ambisonic audio was investigated in Kostrova and Yuri-Andersson (2020) to find if users feel less prone to motion sickness in the presence of binaural ambisonic audio. Their results found that users feel less prone to simulator sickness in binaural ambisonic audio as compared to stereo audio.

There are a number of factors that are associated with videos such as resolution, motion, frame rate, and projection scheme that may have an impact on simulator sickness. In Chap. 6, the impact of resolution, bitrate, motion in video sequences (temporal information), test session duration, and Motion-to-High-Resolution Latency on the scores of simulator sickness is investigated.

2.5.2 *Evaluation of Simulator Sickness*

The extent of simulator sickness symptoms experienced by a user can be measured by using objective and subjective methods (Weech et al. 2019). Objective measures include measurement of physiological signals such as heart rate, galvanic skin response, and respiration (Katsigiannis et al. 2019; Dennison et al. 2016; Geršak et al. 2020). In the following section, an overview of questionnaires that are being used for assessing simulator sickness among users will be provided to facilitate the researchers in selecting an appropriate test method according to their needs. Table 2.2, which is adapted from Chang et al. (2020), lists the possible long questionnaires to measure simulator sickness.

Table 2.2 List of simulator sickness questionnaires. It is adapted from Chang et al. (2020)

S. no	Authors	Year	Name of questionnaire	Number of symptoms	Scale	Number of factors
1	Kennedy et al. (1993)	1993	SSQ	16	4-point	3
2	Gianaros et al. (2001)	2001	MSAQ	16	9-point	4
3	Ames et al. (2005)	2005	VRSQ	13	7-point	2
4	Bouchard et al. (2007)	2007	F-SSQ	16	4-point	2
5	Stone (2017)	2017	CSQ	9	4-point	2
6	Kim et al. (2018)	2018	VRSQ	9	4-point	2

The most popular measurement technique to assess the symptoms of simulator sickness is the SSQ proposed by Kennedy et al. (1993). This questionnaire contains 16 symptoms, and users rate each symptom using a four-point scale. They categorized these symptoms into three clusters, namely, Nausea (N), Oculomotor (O), and Disorientation (D). Motion Sickness Assessment Questionnaire (MASQ) is proposed by Gianaros et al. (2001), where 16 symptoms/descriptors are categorized into 4 categories (Gastrointestinal, Central, Peripheral, Sopite-related) and each descriptor is rated on a 9-point scale. Ames et al. (2005) proposed a Virtual Reality Symptom Questionnaire (VRSQ) that contains 13 symptoms, and users rate each symptom using a 7-category rating scale. The authors categorized these symptoms in two-factor structure namely, Ocular and Nonocular. Bouchard et al. (2007) re-examined the factor structure of the SSQ, and based on their results, they proposed the two-factor structure (Nausea and Oculomotor) instead of the three-factor structure for the SSQ without removing any of the symptoms from the SSQ.

Using the SSQ Kennedy et al. (1993), some researchers made an attempt to reduce the number of symptoms asked from participants during the test. Stone (2017) retained only 9 symptoms out of 16, after a reduction via Principal Component Analysis (PCA). Furthermore, they categorized these symptoms into a two-factor structure Dizziness and Difficulty in Focusing. Kim et al. (2018) derived the Virtual Reality Sickness Questionnaire (VRSQ) using PCA and reduced the number of symptoms from 16 to 9. They categorized these 9 symptoms into a two-factor structure (Oculomotor and Disorientation).

There are several single-scale questions that are used to assess the well-being of users using a Likert scale (Wertheim et al. 2001; Keshavarz and Hecht 2011; Tran et al. 2017; Perez et al. 2018; Raake et al. 2020). The Misery Scale (MISC) was proposed by Wertheim et al. (2001), and it uses a 0–10 scale (0: No problems, 1: Uneasy feeling in the head ... 10: Vomiting). The Fast Motion Sickness (FMS) was developed by Keshavarz and Hecht (2011) and uses a range from 0 to 20 (0: No sickness at all, ... 20: Frank sickness, i.e., imminent vomiting). Short SSQ was proposed by Tran et al. (2017) that assesses the level of nausea or dizziness on a 5-point scale (1: Very dizzy, 2: Dizzy, 3: Slightly dizzy, 4: Not dizzy, and 5: Absolutely not dizzy). Vertigo scale was proposed by Perez et al. (2018) to report sickness scores using a 5-point scale (1: Unbearable, 2: Unpleasant, 3: Uncomfortable, 4: Light effects, and 5: No problem). Raake et al. (2020) framed the discomfort question by asking about the viewing comfort of users according to ITU-R BT. 2021 to promote a positive outlook.

The approaches that have been reviewed here can give an idea to researchers on the test method to use for their assessment. Further, all test methods use a discrete scale to rate sickness via different rating schemes (0–3, 0–6, 0–10, 0–20, 1–5). The main limitation of the single-scale question is that it cannot fully record the symptoms of participants. But the problem with the full questionnaires is that users may not self-report the symptoms in an accurate manner, as they may get bored.

A pre-test questionnaire could be used to know about the susceptibility of users to motion sickness using a motion sickness susceptibility questionnaire (Golding 1998).

The SSQ was adapted for VR contents by Kim et al. (2018) and Stone (2017), but not for 360° contents. 360° contents differ in terms of DoF and the way they are captured. Here, VR video contains mainly computer-generated imagery and can provide movement to a user in 6 DoF (3 directional and 3 rotational). On the other hand, 360° video is captured from real-life scenes using a 360° camera and can provide movement to a user in only 3 DoF (3 rotational). Hence, it could be possible that users may feel less prone to simulator sickness in 360° videos than in VR. Therefore, it would be reasonable to investigate if it is efficient to use the SSQ in its current form for 360° videos (Stone 2017; Rebenitsch and Owen 2016), which is addressed in Chap. 7. Furthermore, the suitability of the single-scale version of the simulator sickness in comparison to the long version of the SSQ is investigated in Chap. 8.

2.6 Evaluation of Presence

Omnidirectional videos allow users to interactively explore the scene, which gives them a perception of being present in the virtual environment (VE). There are several questionnaires that are being used to assess presence (Witmer and Singer 1998; Usoh et al. 2000; Lessiter et al. 2001). Table 2.3 lists the possible long questionnaires to measure presence.

Table 2.3 List of presence questionnaires

S. no	Authors	Year	Name of questionnaire	Number of items	Scale	Number of factors
1	Witmer and Singer (1998)	1998	PQ	32	7-point	4
2	Witmer and Singer (1998)	1998	ITQ	29	7-point	3
3	Lessiter et al. (2001)	2001	ITC-SOPI	44	5-point	4
4	Usoh et al. (2000)	2000	SUS	6	7-point	N/A
5	igroup[a]	–	IPQ	14	7-point	4

[a] http://www.igroup.org/pq/ipq/download.php

To measure presence in a VE, the presence questionnaire (PQ) (Witmer and Singer 1998) has been widely used. The PQ (version 2) consists of 32 questions that are further classified into four different categories: control, sensory, distraction, and realism factors. Subjects rate each factor on a 7-point scale, with different descriptive labels for some factors. To measure the tendency of immersion, the Immersive Tendencies Questionnaire (ITQ) is used (Witmer and Singer 1998). This questionnaire contains 29 items that are classified into 3 factors: involvement, focus, and tendency to play games. Each item has to be rated on a 7-point scale. The ITC-Sense of Presence Inventory (ITC-SOPI) is proposed by Lessiter et al. (2001) to measure presence. Their questionnaire contains a relatively larger number of questions (44) and is factored into four factors: Sense of Physical Space, Engagement, Ecological Validity and Negative Effects. A 5-point Likert scale is used to rate each question. The Igroup Presence Questionnaire (IPQ)[5] contains 14 items that are sub-divided into four factors: General Presence, Spatial Presence, Involvement, and Experienced Realism. Subjects need to rate each question on a 7-point scale. A relatively short questionnaire, SUS (Slater-Usoh-Steed), that contains only six questions is proposed in Usoh et al. (2000). These six questions were not categorized further. To rate each question, a 7-point scale is used with different descriptive labels.

Similarly to the case of the SSQ, short versions of the approach have also been applied. For example, the Short Presence Questionnaire (Short PQ) as described in Tran et al. (2017) can be used. For the Short PQ, subjects are asked to answer one question on a 5-point scale (1: absolutely no sense of presence and 5: a true sense of presence as in a real environment).

Several studies have evaluated presence and simulator sickness simultaneously (Jerome et al. 2005; Mur-Spiegl and Schlögl 2018; Weech et al. 2019). For example, in Jerome et al. (2005), Jerome et al. explored the relationships between presence, simulator sickness, and length of the simulator exposure. Experimental results showed that the duration of exposure is positively correlated with simulator sickness and presence; however, simulator sickness and presence are negatively correlated. Furthermore, Mur-Spiegl and Schlögl (2018) compared simulator sickness and presence in low-quality and high-quality HMDs. Results showed that a higher quality HMD has a positive effect on the felt presence. But their results are inconclusive with respect to the effect of HMDs on simulator sickness. In Weech et al. (2019), Weech et al. concluded that simulator sickness and presence have a negative relationship. Seay et al. (2001) measured the effect of FOV on simulator sickness and presence. The authors found that higher presence was reported in the high FOV than that of low FOV conditions. Also, a higher FOV leads to higher nausea scores as compared to a lower FOV. However, the relation between presence and simulator sickness could not be deduced. In Ling et al. (2013), Ling et al. found that presence and simulator sickness are positively correlated.

[5] http://www.igroup.org/pq/ipq/download.php.

Unfortunately, due to the lack of comprehensive studies in the state of the art, the nature of the relationship between simulator sickness and presence is hard to reveal. In Chap. 8, an attempt is made to derive the relation between these two. Additionally, the well-established PQ method is compared with only one single scale, to analyze how well presence can be captured using only a single scale.

2.7 Exploration Behavior

When consuming videos using traditional 2D displays, the user is restricted to watch only a certain viewport of the scene. However, in omnidirectional videos, viewers can look wherever they want. The users become the director of the movie and are free to watch any view. With this freedom, a lot of researchers are interested to find out at which specific parts of the video users are looking at any given moment in time. In other words, they are interested in investigating the exploration behavior of the users. In addition to the head rotation data is to track eye movements while exploring 360° videos (David et al. 2018). For the present book, only head tracking will be focused on.

There are several works reported in the literature on the assessment of the exploration behavior of users (Corbillon et al. 2017b; Rai et al. 2017; Curcio et al. 2017; Rossi et al. 2020; Chao et al. 2020). Corbillon et al. (2017b) provided a dataset that includes the head rotation data for 59 users. They introduced several methods to analyze the head rotation data. They analyzed the maximum angular distance from the initial head position (at 0 s) for different durations of time (1 s, 2 s, 3 s, and 5 s) for each user and each video. Further, they analyzed the probability of a pixel for a user during the whole video duration. Rai et al. (2017) provided guidelines to assess and compare the saliency maps for 360° images. The Linear Correlation Coefficient metric is used to compare the saliency maps. Further, they compared head and head–eye saliency maps and found that these two saliency maps correlate poorly. The distribution of fixations in the yaw direction was shown by Chao et al. (2020). Further, they found that visual attention is concentrated on salient regions in the presence of sound, and it scatters widely when there is no sound while viewing omnidirectional videos. Curcio et al. (2017) proposed a new metric called Similarity Ring Metric (SRM) that measures the degree of similarity between viewing behavior for one subject or between several subjects. In this book, the impact of exploration behavior is investigated on video quality (Sect. 5.3.6) and simulator sickness Sect. 6.2 scores.

There are publicly available tools that can be used to conduct subjective tests using HMDs and collect the corresponding head rotation data, such as AVTrack360 (Fremerey et al. 2018), MIRO360 (Pérez and Escobar 2019), and VRate (Regal et al. 2018).

For example, AVTrack (Fremerey et al. 2018) is a Python-based framework for capturing the exploration behavior of people while watching 360° videos. It uses the Whirligig player for playing out 360° videos. Together with the tool, the authors published a dataset providing the head rotation data for 48 subjects watching 20 different omnidirectional videos with 30 s duration in a task-free scenario. Four different methods were proposed to evaluate the head rotation data: the percentage of subjects discovering the respective parts of the video, the percentage of time spent by the subjects watching specific areas of 360° video, heatmaps, and angular yaw speed heatmaps. Heatmaps are found to be a good way for visualizing how people are exploring omnidirectional videos, while angular yaw speed heatmaps can be used to investigate the effect of specific video events at a specific timestamp on the exploration behavior.

MIRO360 (Pérez and Escobar 2019) is an Android-based tool that can be used to carry out the subjective test with DSIS and ACR test methods using Samsung Gear VR. This tool also records the head rotation data of users while watching the 360° videos. Additionally, after the video is over, there are different questions that can be asked from users related to quality, simulator sickness, and presence, while their responses can be recorded using a controller.

VRate (Regal et al. 2018) was developed using Unity3D that allows the researchers to ask self-reported questionnaires from the test participants without breaking them from a virtual environment. However, it is not clear if this tool records the head rotation data of users.

The publicly available tools are not capable of playing 360° video and VR together with ambisonics audio. Please note that no work is continued in this direction.

References

A. Abbas, B. Adsumilli, *AHG8: New GoPro Test Sequences for Virtual Reality Video Coding*. JVET-D0026, 4th Meeting: Chengdu, China, 15–21 October, (2016)

P.R. Alface, J. Macq, N. Verzijp, Interactive omnidirectional video delivery: a bandwidth-effective approach. Bell Labs Tech. J. **16**(4), 135–147 (2012). ISSN: 1538-7305. https://doi.org/10.1002/bltj.20538

S.L. Ames, J. Wolffsohn, N. Mcbrien, The development of a symptom questionnaire for assessing virtual reality viewing using a head-mounted display. Optom. Vis. Sci. **82**, 168–176 (2005)

M.S. Anwar et al., Subjective QoE of 360-degree virtual reality videos and machine learning predictions. IEEE Access **8**, 148084–148099 (2020). https://doi.org/10.1109/ACCESS.2020.3015556

K.W. Arthur, Effects of Field of View on Performance with Head-mounted Displays. AAI9968542. PhD Thesis, 2000. ISBN: 0-599-73372-1

E. Asbun et al., *AHG8: InterDigital Test Sequences for Virtual Reality Video Coding*. JVET-D0039, 4th Meeting: Chengdu, China, 15–21 October (2016)

Y. Bao et al., Shooting a moving target: Motion-prediction-based transmission for 360-degree videos, in *2016 IEEE International Conference on Big Data (Big Data)* (2016), pp. 1161–1170. https://doi.org/10.1109/BigData.2016.7840720

K.R. Boff, J.E. Lincoln, *Engineering Data Compendium: Human Perception and Performance*. AAMRL Wright-Patterson AFB, OH (1988)

S. Bouchard, G. Robillard, P. Renaud, Revising the factor structure of the simulator sickness questionnaire. Summer. Annu. Rev. Cybertherapy Telemed. **5**, 128–137 (2007)
K. Brunnström et al., Latency impact on quality of experience in a virtual reality simulator for remote control of machines. Signal Process. Image Commun. **89**, 116005 (2020). ISSN: 0923-5965. https://doi.org/10.1016/j.image.2020.116005. https://www.sciencedirect.com/science/article/pii/S0923596520301648
E. Chang, H. Kim, B. Yoo, Virtual reality sickness: a review of causes and measurements. Int. J. Human-Comput. Interact. **36**, 1658–1682 (2020)
F.-Y. Chao et al., Audio-visual perception of omnidirectional video for virtual reality applications, in *2020 IEEE International Conference on Multimedia Expo Workshops (ICMEW)* (2020), pp. 1–6. https://doi.org/10.1109/ICMEW46912.2020.9105956
X. Corbillon et al., Viewport-adaptive navigable 360-degree video delivery, in *2017 IEEE International Conference on Communications (ICC)* (2017a), pp. 1–7. ISSN: 1938-1883. https://doi.org/10.1109/ICC.2017.7996611
X. Corbillon, F. De Simone, G. Simon, 360-degree video head movement dataset, in *Proceedings of the 8th ACM on Multimedia Systems Conference*. MMSys'17 (Association for Computing Machinery, Taipei, 2017b), pp. 199–204. ISBN: 9781450350020. https://doi.org/10.1145/3083187.3083215
C. Cortés et al., Influence of video delay on quality, presence, and sickness in viewport adaptive immersive streaming, in *2020 Twelfth International Conference on Quality of Multimedia Experience (QoMEX)* (2020), pp. 1–4. https://doi.org/10.1109/QoMEX48832.2020.9123114
A. Covaci et al., 360-degree mulsemedia: A way to improve subjective QoE in 360-degree videos, in *Proceedings of the 27th ACM International Conference on Multimedia*. MM '19 (Association for Computing Machinery, Nice, 2019), pp. 2378–2386. ISBN: 9781450368896. https://doi.org/10.1145/3343031.3350954
I.D.D. Curcio, H. Toukomaa, D. Naik, 360-degree video streaming and its subjective quality, in *SMPTE 2017 Annual Technical Conference and Exhibition* (2017), pp. 1–23. https://doi.org/10.5594/M001758
I.D.D. Curcio, H. Toukomaa, D. Naik, Bandwidth reduction of omnidirectional viewport-dependent video streaming via subjective quality assessment, in *Proceedings of the 2nd International Workshop on Multimedia Alternate Realities*. AltMM '17 (ACM, Mountain View, 2017), pp. 9–14. ISBN: 978-1-4503-5507-0. http://doi.acm.org/10.1145/3132361.3132364
E.J. David et al., A dataset of head and eye movements for 360 videos, in *Proceedings of the 9th ACM Multimedia Systems Conference*. MMSys '18. Amsterdam, Netherlands (2018), pp. 432–437. ISBN: 9781450351928
M.S. Dennison, A.Z. Wisti, M. D'Zmura, Use of physiological signals to predict cybersickness. Displays **44**, 42–52 (2016). Contains Special Issue Articles – Proceedings of the 4th Symposium on Liquid Crystal Photonics (SLCP 2015). ISSN: 0141-9382. https://doi.org/10.1016/j.displa.2016.07.002. http://www.sciencedirect.com/science/article/pii/S0141938216301081
P. DiZio, J. Lackner, Circumventing side effects of immersive virtual environments, in *HCI* (1997)
H. Duan et al., IVQAD 2017: An immersive video quality assessment database, in *2017 International Conference on Systems, Signals and Image Processing (IWSSIP)* (2017), pp. 1–5. https://doi.org/10.1109/IWSSIP.2017.7965610
H.B.-L. Duh et al., Effects of field of view on balance in an immersive environment, in *Proceedings IEEE Virtual Reality 2001* (2001), pp. 235–240. https://doi.org/10.1109/VR.2001.913791
H.B.-L. Duh et al., "Conflicting" motion cues to the visual and vestibular self-motion systems around 0.06 Hz evoke simulator sickness. Human Fact. **46**(1), 142–153 (2004). PMID: 15151161. https://doi.org/10.1518/hfes.46.1.142.30384
N. Dużmańska, P. Strojny, A. Strojny, Can simulator sickness be avoided? A review on temporal aspects of simulator sickness. Front. Psychol. **9**, 2132 (2018)
C.-L. Fan et al., A survey on 360° video streaming: acquisition, transmission, and display. ACM Comput. Surv. **52**(4), 1–36 (2019). ISSN: 0360-0300. https://doi.org/10.1145/3329119

S. Fremerey et al., AVtrack360: An open dataset and software recording people's head rotations watching 360-degree; videos on an HMD, in *Proceedings of the 9th ACM MMSys*. Amsterdam, Netherlands (2018). ISBN: 978-1-4503-5192-8. https://doi.org/10.1145/3204949.3208134. http://doi.acm.org/10.1145/3204949.3208134

V.R. Gaddam et al., Tiling in interactive panoramic video: approaches and evaluation. IEEE Trans. Multimedia **18**(9), 1819–1831 (2016). ISSN: 1520-9210. https://doi.org/10.1109/TMM.2016.2586304

G. Geršak, H. Lu, J. Guna, Effect of VR technology matureness on VR sickness. Multimedia Tools Appl. **79**(21–22), 14491–14507 (2020). ISSN: 1380-7501. https://doi.org/10.1007/s11042-018-6969-2

A. Ghosh, V. Aggarwal, F. Qian, A rate adaptation algorithm for tile-based 360-degree video streaming. CoRR **abs/1704.08215** (2017). arXiv: `1704.08215`. http://arxiv.org/abs/1704.08215

P.J. Gianaros et al., A questionnaire for the assessment of the multiple dimensions of motion sickness. Aviation Space Environ. Med. **72**(2), 115-119 (2001)

J.F. Golding, Motion sickness susceptibility questionnaire revised and its relationship to other forms of sickness. Brain Res. Bull. **47**(5), 507–516 (1998). ISSN: 0361-9230. https://doi.org/10.1016/S0361-9230(98)00091-4. http://www.sciencedirect.com/science/article/pii/S0361923098000914

M. Graf, C. Timmerer, C. Mueller, Towards bandwidth efficient adaptive streaming of omnidirectional video over HTTP: Design, implementation, and evaluation, in *Proceedings of the 8th ACM on Multimedia Systems Conference*. MMSys'17 (2017), pp. 261–271. https://doi.org/10.1145/3083187.3084016. http://doi.acm.org/10.1145/3083187.3084016

J. Guna et al., Influence of video content type on users' virtual reality sickness perception and physiological response. Future Gener. Comput. Syst. **91**, 263–276 (2019)

J. Gutierrez et al., Subjective evaluation of visual quality and simulator sickness of short 360° videos: ITU-T Rec. P.919. IEEE Trans. Multimedia **24**, 3087–3100 (2022). https://doi.org/10.1109/TMM.2021.3093717

K.S. Hale, K.M. Stanney, *Handbook of Virtual Environments: Design, Implementation, and Applications* (CRC Press, Boca Raton, 2014)

P. Hanhart et al., 360-degree video quality evaluation, in *2018 Picture Coding Symposium (PCS)* (2018), pp. 328–332. https://doi.org/10.1109/PCS.2018.8456255

F. Hofmeyer et al., Impacts of internal HMD playback processing on subjective quality perception. Electron. Imag. **2019**(12), 219-1 (2019)

M. Hosseini, V. Swaminathan, Adaptive 360 (VR) video streaming: divide and conquer! CoRR **abs/1609.08729** (2016). arXiv: `1609.08729`. http://arxiv.org/abs/1609.08729

ITU-R BT.500-13, *Recommendation ITU-R BT.500-13: Methodology for the Subjective Assessment of the Quality of Television Pictures* (2012)

ITU–T Rec. P.10/G.100, *Vocabulary for Performance and Quality of Service* (International Telecommunication Union, Geneva, 2017)

C. Jerome et al., The effects of presence and time of exposure on simulator sickness, in *Proceedings of the Human Factors and Ergonomics Society Annual Meeting*. vol. 49, no. 26 (SAGE Publications Sage CA, Los Angeles, 2005), pp. 2258–2262

D.M. Johnson, Introduction to and review of simulator sickness research. US Army Research Institute for the Behavioral and Social Sciences (Rotary-Wing Aviation Research Unit, Fort Rucke, 2005)

S. Katsigiannis, R. Willis, N. Ramzan, A QoE and simulator sickness evaluation of a smart-exercise-bike virtual reality system via user feedback and physiological signals. IEEE Trans. Consum. Electron. **65**(1), 119–127 (2019)

R.S. Kennedy et al., Simulator sickness questionnaire: an enhanced method for quantifying simulator sickness. Int. J. Aviat. Psychol. **3**(3), 203–220 (1993)

R.S. Kennedy et al., Configural scoring of simulator sickness, cyber-sickness and space adaptation syndrome: Similarities and differences, in *Virtual and Adaptive Environments: Applications, Implications, and Human Performance Issues* (Lawrence Erlbaum Associates Publishers, Mahwah, 2003), p. 247

B. Keshavarz, H. Hecht, Validating an efficient method to quantify motion sickness. Human Factors **53**(4), 415–426 (2011). PMID: 21901938. https://doi.org/10.1177/0018720811403736

H.K. Kim et al., Virtual reality sickness questionnaire (VRSQ): motion sickness measurement index in a virtual reality environment. Appl. Ergon. **69**, 66–73 (2018). ISSN: 0003-6870. https://doi.org/10.1016/j.apergo.2017.12.016. http://www.sciencedirect.com/science/article/pii/S000368701730282X

E.M. Kolasinski, *Simulator Sickness in Virtual Environments*. US Army Research Institute for the Behavioral and Social Sciences, vol. 1027 (IEEE, Piscataway, 1995)

M. Kostrova, V. Yuri-Andersson. *Designing with Sound to Reduce Motion Sickness in VR*. Bachelor Thesis, Uppsala University, 2020

J.J. LaViola Jr., A discussion of cybersickness in virtual environments. SIGCHI Bull. **32**(1), 47–56 (2000). ISSN: 0736-6906. https://doi.org/10.1145/333329.333344. http://doi.acm.org/10.1145/333329.333344

P. Le Callet, S. Möller, A. Perkis et al., Qualinet white paper on definitions of quality of experience, in *European Network on Quality of Experience in Multimedia Systems and Services (COST Action IC 1003)*, 3.2012 (2012)

J. Lessiter et al., A cross-media presence questionnaire: the ITC-sense of presence inventory. Presence **10**(3), 282–297 (2001). https://doi.org/10.1162/105474601300343612

B.J. Li et al., A public database of immersive VR videos with corresponding ratings of arousal, valence, and correlations between head movements and self report measures. Front. Psychol. **8**, 2116 (2017). ISSN: 1664-1078. https://doi.org/10.3389/fpsyg.2017.02116. https://www.frontiersin.org/article/10.3389/fpsyg.2017.02116

C. Li et al., Bridge the gap between VQA and human behavior on omnidirectional video: A large-scale dataset and a deep learning model, in *Proceedings of the 26th ACM International Conference on Multimedia*. MM '18 (Association for Computing Machinery, Seoul, 2018), pp. 932–940. ISBN: 9781450356657. https://doi.org/10.1145/3240508.3240581

Y. Ling et al., The relationship between individual characteristics and experienced presence. Comput. Hum. Behav. **29**(4), 1519–1530 (2013). ISSN: 0747-5632. https://doi.org/10.1016/j.chb.2012.12.010

W.-C. Lo et al., 360-degree video viewing dataset in head-mounted virtual reality, in *Proceedings of the 8th ACM on Multimedia Systems Conference*. MMSys'17 (Association for Computing Machinery, Taipei, 2017a), pp. 211–216. ISBN: 9781450350020. https://doi.org/10.1145/3083187.3083219

W.-C. Lo et al., Performance measurements of 360° video streaming to head-mounted displays over live 4G cellular networks, in *2017 19th Asia-Pacific Network Operations and Management Symposium (APNOMS)* (2017b), pp. 205–210. https://doi.org/10.1109/APNOMS.2017.8094203

F. Lopes et al., Subjective and objective quality assessment of omnidirectional video, in *SPIE Conference Series* (2018). https://doi.org/10.1117/12.2321679

S. Mahmoudpour, P. Schelkens, Visual quality analysis of judder effect on head mounted displays, in *2019 27th European Signal Processing Conference (EUSIPCO)* (2019), pp. 1–5. https://doi.org/10.23919/EUSIPCO.2019.8902665

K. Mania et al., Perceptual sensitivity to head tracking latency in virtual environments with varying degrees of scene complexity, in *Proceedings of the 1st Symposium on Applied Perception in Graphics and Visualization*. APGV '04 (ACM, Los Angeles, 2004), pp. 39–47. ISBN: 1-58113-914-4. https://doi.org/10.1145/1012551.1012559. http://doi.acm.org/10.1145/1012551.1012559

C.A. Maxwell, Flicker science and the consumer. Informat. Display **8**, 7 (1992)

K. Misra et al., An overview of tiles in HEVC. IEEE J. Select. Top. Signal Process. **7**(6), 969–977 (2013). ISSN: 1932-4553. https://doi.org/10.1109/JSTSP.2013.2271451

R. Monnier, R. van Brandenburg, R. Koenen, Streaming UHD-quality VR at realistic bitrates: mission impossible? in *2017 NAB Broadcast Engineering and Information Technology Conference (BEITC)* (2017), pp. 1–8

D. Mur-Spiegl, S. Schlögl, Simulator sickness and felt presence: comparing low-and high-end virtual reality headsets, in *Proceedings of the Forschungsforum der Osterreichischen Fachhochschulen* (2018), pp. 1–6

A.T. Nasrabadi et al., A taxonomy and dataset for 360 & deg; videos, in *Proceedings of the 10th ACM Multimedia Systems Conference*. MMSys '19 (ACM, Amherst, 2019), pp. 273–278. ISBN: 978-1-4503-6297-9. https://doi.org/10.1145/3304109.3325812. http://doi.acm.org/10.1145/3304109.3325812

R. Pausch, T. Crea, M. Conway, A literature survey for virtual environments: military flight simulator visual systems and simulator sickness. Presenc. Teleoper. Virtual Environ. **1**(3), 344–363 (1992). ISSN: 1054-7460. https://doi.org/10.1162/pres.1992.1.3.344

P. Pérez, J. Escobar, MIRO360: A tool for subjective assessment of 360 degree video for ITU-T P.360-VR, in *2019 Eleventh International Conference on Quality of Multimedia Experience (QoMEX)* (2019), pp. 1–3. https://doi.org/10.1109/QoMEX.2019.8743216

P. Perez et al., Towards systematic analysis of cybersickness in high motion omnidirectional video, in *International Conference on Quality of Multimedia Experience*. Sardinia, Italy (2018)

S. Petrangeli et al., An HTTP/2-based adaptive streaming framework for 360-degree virtual reality videos, in *Proceedings of the 2017 ACM on Multimedia Conference*. MM '17 (ACM, Mountain View, 2017), pp. 306–314. ISBN: 978-1-4503-4906-2. https://doi.org/10.1145/3123266.3123453. http://doi.acm.org/10.1145/3123266.3123453

T. Philipp, *How Spatial Audio affects Motion Sickness in Virtual Reality*. Bachelor Thesis, Hochschule der Medien, 2020

M.H. Pinson, S. Wolf, A new standardized method for objectively measuring video quality. IEEE Trans. Broadcast. **50**(3), 312–322 (2004). ISSN: 1557-9611. https://doi.org/10.1109/TBC.2004.834028

A. Raake et al., SiSiMo: Towards simulator sickness modeling for 360-degree videos viewed with an HMD, in *2020 IEEE Conference on Virtual Reality and 3D User Interfaces Abstracts and Workshops (VRW)* (2020), pp. 582–583

Y. Rai, J. Gutiérrez, P. Le Callet, A dataset of head and eye movements for 360 degree images, in *Proceedings of the 8th ACM on Multimedia Systems Conference*. MMSys'17 (Association for Computing Machinery, Taipei, 2017), pp. 205–210. ISBN: 9781450350020. https://doi.org/10.1145/3083187.3083218

J.T. Reason, J.J. Brand, *Motion Sickness* (Academic, London, 1975)

L. Rebenitsch, C. Owen, Review on cybersickness in applications and visual displays. Virtual Real. **20**(2), 101–125 (2016). ISSN: 1359-4338. https://doi.org/10.1007/s10055-016-0285-9

L. Rebenitsch, C. Owen, *Estimating Cybersickness from Virtual Reality Applications*, vol. 25(1) (Springer, Berlin, Heidelberg, 2021), pp. 165–174. https://doi.org/10.1007/s10055-020-00446-6. ISSN: 1359-4338

G. Regal et al., VRate: A unity3D asset for integrating subjective assessment questionnaires in virtual environments, in *2018 Tenth International Conference on Quality of Multimedia Experience (QoMEX)* (2018), pp. 1–3

G.E. Riccio, T.A. Stoffregen, An ecological theory of motion sickness and postural instability. Ecol. Psychol. **3**(3), 195–240 (1991). https://doi.org/10.1207/s15326969eco0303_2

S. Rossi et al., Do users behave similarly in VR? Investigation of the user influence on the system design. ACM Trans. Multimedia Comput. Commun. Appl. **16**(2) (2020). ISSN: 1551-6857. https://doi.org/10.1145/3381846

R. Schatz et al., Towards subjective quality of experience assessment for omnidirectional video streaming, in *In 2017 Ninth International Conference on Quality of Multimedia Experience (QoMEX)* (2017), pp. 1–6. ISSN: 2472-7814. https://doi.org/10.1109/QoMEX.2017.7965657

S. Schwarz, A. Aminlou et al., *Tampere Pole Vaulting Sequence for Virtual Reality Video Coding*. JVET-D0143, 4th Meeting: Chengdu, China, 15–21 October (2016)

A.F. Seay et al., Simulator sickness and presence in a high FOV virtual environment, in *Proceedings IEEE Virtual Reality 2001* (2001), pp. 299–300. https://doi.org/10.1109/VR.2001.913806

A. Singla et al., Quality assessment protocols for omnidirectional video quality evaluation. Electron. Imag. **2020**(11), 69-1–69-7 (2020). ISSN: 2470-1173. https://doi.org/10.2352/ISSN.2470-1173.2020.11.HVEI-069. https://www.ingentaconnect.com/content/ist/ei/2020/00002020/00000011/art00003

B.J. Sitter et al., 78-3: Screen door effect reduction with diffractive film for virtual reality and augmented reality displays, in *SID Symposium Digest of Technical Papers* (2017)

M. Slater, A note on presence terminology. Presence Connect **3**, 1–5 (2003)

R.M. Stern et al., The effects of fixation and restricted visual field on vection-induced motion sickness. Aviation Space Environ. Med. **61**(8), 712–715 (1990)

W.B. Stone III, *Psychometric Evaluation of the Simulator Sickness Questionnaire as a Measure of Cybersickness*. Doctoral Dissertation, Iowa State University (2017)

G.J. Sullivan et al., Overview of the high efficiency video coding (HEVC) standard. IEEE Trans. Circuit. Syst. Video Technol. **22**(12), 1649–1668 (2012). ISSN: 1051-8215. https://doi.org/10.1109/TCSVT.2012.2221191

W. Sun, R. Guo. *Test Sequences for Virtual Reality Video Coding from LetinVR*. JVET-D0179, 4th Meeting: Chengdu, China, 15–21 October (2016)

H.T.T. Tran et al., A subjective study on QoE of 360 video for VR communication, in *IEEE 19th International Workshop on MMSP* (2017), pp. 1–6

M. Treisman, Motion sickness: an evolutionary hypothesis. Science **197**(4302), 493-495 (1977)

E. Upenik, M. Rerabek, T. Ebrahimi, Testbed for subjective evaluation of omnidirectional visual content, in *2016 Picture Coding Symposium (PCS)* (2016), 1–5. https://doi.org/10.1109/PCS.2016.7906378

M. Usoh et al., Using presence questionnaires in reality. Presence Teleoper. Virtual Environ. **9**(5), 497–503 (2000)

R. van Brandenburg, R. Koenen, D. Sztykman, *CDN Optimisation for VR Streaming* (2017). https://www.ibc.org/tech-advances/cdn-optimisation-for-vr-streaming-/2457. Article (12 Jun 2018)

A. van Kasteren et al., Quality of experience assessment of 360-degree video. Electron. Imag. **2020**(11), 91-1–91-7 (2020). ISSN: 2470-1173. https://doi.org/10.2352/ISSN.2470-1173.2020.11.HVEI-091. https://www.ingentaconnect.com/content/ist/ei/2020/00002020/00000011/art00004

S. Weech, S. Kenny, M. Barnett-Cowan, Presence and cybersickness in virtual reality are negatively related: a review. Front. Psychol. **10**, 158 (2019)

A.H. Wertheim, J.E. Bos, A.J. Krul, *Predicting Motion Induced Vomiting from Subjective Misery (MISC) Ratings Obtained in 12 Experimental Studies*. Report TNO-TM-01-A066 (TNO Human Factors Research Institute, Soesterberg, 2001)

M.L. Wilson, The effect of varying latency in a head-mounted display on task performance and motion sickness. Doctoral Dissertation (2016)

B.G. Witmer, M.J. Singer, Measuring presence in virtual environments: a presence questionnaire. Presence **7**(3), 225–240 (1998)

M. Xu et al., A subjective visual quality assessment method of panoramic videos, in *2017 IEEE International Conference on Multimedia and Expo (ICME)* (2017), pp. 517–522. https://doi.org/10.1109/ICME.2017.8019351

S.-H. Yao, C.-L. Fan, C.-H. Hsu, Towards quality-of-experience models for watching 360° videos in head-mounted virtual reality, in *2019 Eleventh International Conference on Quality of Multimedia Experience (QoMEX)* (2019), pp. 1–3. https://doi.org/10.1109/QoMEX.2019.8743198

S.-C. Yen, C.-L. Fan, C.-H. Hsu, Streaming 360° videos to head-mounted virtual reality using DASH over QUIC transport protocol, in *Proceedings of the 24th ACM Workshop on Packet Video* (Association for Computing Machinery, New York, 2019), pp. 7–12. ISBN: 9781450363006. https://doi.org/10.1145/3304114.3325616

A. Zare et al., HEVC-compliant tile-based streaming of panoramic video for virtual reality applications, in *Proceedings of the 2016 ACM on Multimedia Conference*. MM '16 (ACM, Amsterdam, 2016), pp. 601–605. ISBN: 978-1-4503-3603-1. https://doi.org/10.1145/2964284.2967292. http://doi.acm.org/10.1145/2964284.2967292

B. Zhang et al., Subjective and objective quality assessment of panoramic videos in virtual reality environments, in *2017 IEEE International Conference on Multimedia Expo Workshops (ICMEW)* (2017), pp. 163–168. https://doi.org/10.1109/ICMEW.2017.8026226

Y. Zhang et al., Subjective panoramic video quality assessment database for coding applications. IEEE Trans. Broadcast. **64**, 461–473 (2018). ISSN: 0018-9316. https://doi.org/10.1109/TBC.2018.2811627

Chapter 3
Technical Setup

3.1 Overview of End-to-End Chain

The end-to-end chain of a 360° video processing and the testing system as used in a subjective test is illustrated in Fig. 3.1. First, source sequences in an appropriate projection scheme, e.g., ERP, have to be identified based on the goal of the study being conducted. These source sequences can be analyzed in terms of spatial and temporal complexities or categorized subjectively based on motion, indoor, outdoor, etc., to ensure the usage of source sequences covering a wide range of complexities (P.910, 2008). Following this, suitable test conditions are selected, and the source sequences are processed using these test conditions. These processed video sequences are referred to as PVS. Finally, the PVSs are presented on an HMD using a 360° video player such as a Whirligig player. The head rotation of a person using the HMD is constantly tracked by the corresponding HMD to correctly display the video in a person's FOV (field of view) or viewport. The viewport is updated based on the head rotation behavior of the user, additionally depending on factors such as rendering delay, motion-to-photon latency, etc. Finally, depending on the goal of the study, the subject/user is asked to perform certain tasks, answer questions, or rate the stimulus or specific percepts on a set of scales. An example of such a user study can be to evaluate the impact of factors such as recording, coding, presentation, etc., on user perception, wherein the subject is asked to assess the quality of the presented audiovisual stimuli and answer additional questionnaires about cybersickness and presence. In addition to answering such questions or performing specific tasks (e.g., describing each video verbally after the video is over), the exploration behavior of the subject can also be analyzed.

A. Singla, *Assessment of Visual Quality and Simulator Sickness for Omnidirectional Videos*, T-Labs Series in Telecommunication Services,
https://doi.org/10.1007/978-3-031-49988-3_3

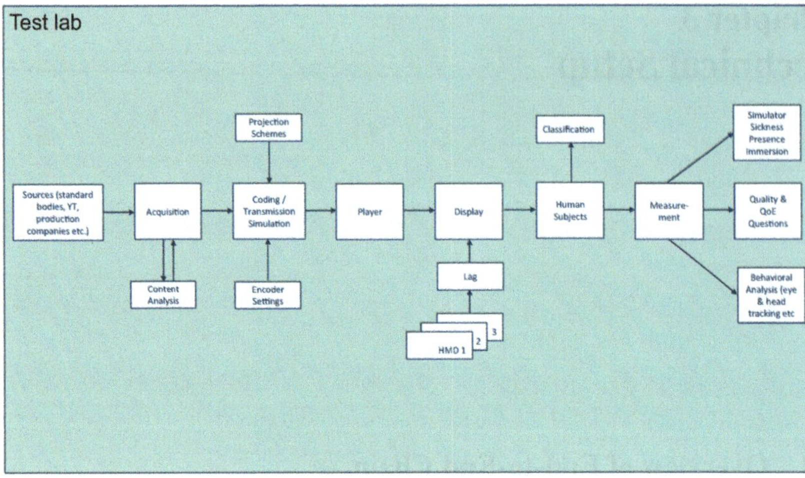

Fig. 3.1 360° video system end-to-end testing environment

3.2 Selection of 360° Videos

During the course of the book, several subjective tests have been conducted, with different datasets. The conducted tests are listed as Test 1, Test 2, Test 3, ... Test 12. These have been published in either peer-reviewed "conferences or journals." Table 3.1 shows a test number, test method, resolution, and frame rate of the videos, HMD used, and its full resolution for Test 1 to Test 12. This section explains the dataset used in the different subjective tests.

3.2.1 Test 1

This section describes Test 1, published in QoMEX 2017 (Singla et al., 2017b). The goal of this test was to evaluate and compare the integral quality and simulator sickness for different omnidirectional contents for Oculus Rift and HTC Vive.

The dataset used for Test 1 was downloaded from YouTube with audio, in two video resolutions (4K and FHD), see Fig. 3.2. The respective resolution has its own individual bitrates (see Table 3.2). All the test sequences are in YUV 420 color space, 8-bits per channel, and ERP format. The reason for downloading the dataset from YouTube was that this test was aimed at longer sequences and the duration of these videos is much longer (60 s) as compared to datasets available at the time from, for example, standardization groups (typically 10 s) (Abbas and Adsumilli, 2016; Asbun et al., 2016; Schwarz et al., 2016; Sun and Guo, 2016). Note that no uncompressed sources were available at the time with appropriate duration. As the videos were already encoded (H.264) with YouTube's pipeline, no further

Table 3.1 Summary of conducted tests to determine test method, various HMDs, video resolution, and frame rate

Test #	Sections	Published at	Method	Resolution	Frame rate	Encoder	HMD name	Total resolution of HMD	No. of subjects	Duration of video
#1. Singla et al. (2017b)	Sect. 3.2.1	QoMEX 2017	ACR	4K, FHD	25–30	H.264/AVC	Oculus Rift,	2160×1200	28	60 s
							HTC Vive	2160×1200		
#2. Singla et al. (2017a)	Sect. 3.2.2	ACM MM 2017	M–ACR	4K, FHD	30	H.265	Oculus Rift	2160×1200	30	10 s
#3. Singla et al. (2018)	Sect. 3.2.2	HVEI 2018	DSIS	4K, FHD	30	HEVC	Oculus Rift	2160×1200	27	10 s
#4. Singla et al. (2019a)[a]	Sect. 3.2.3		ACR	4K, 6K, 8K	30	HEVC	HTC Vive Pro	2880×1600	29	10 s
#5. Singla et al. (2019a)	Sect. 3.2.3	MMSP 2019	M–ACR	4K, 6K, 8K	30	HEVC	HTC Vive Pro	2880×1600	30	10 s
#6. Singla et al. (2019a)	Sect. 3.2.3		DSIS	4K, 6K, 8K	30	HEVC	HTC Vive Pro	2880×1600	28	10 s
#7. Fremerey et al. (2018)	Sect. 3.2.4	ACM MMSys 2018	–	4K	30	H.264	HTC Vive	2160×1200	48	30 s
#8. Singla et al. (2019c)[b]	Sect. 3.2.5	ACM MMSys 2019	ACR	4K, 6K, 8K	25, 60	H.265	Gear VR (S8)	2960×1440	28	30 s
#9. Singla et al. (2019c)	Sect. 3.2.5		ACR	4K, 6K, 8K	25, 60	H.265	Gear VR (S8)	2960×1440	25	30 s
#10. Raake et al. (2020)	Sect. 3.2.6	VR 2020	ACR	4K	30	H.265	HTC Vive Pro	2880×1600	28	60 s
#11. Gutierrez et al. (2022)	Sect. 3.2.7	TMM 2021	ACR	4K	25, 30	H.265	HTC Vive Pro	2880×1600	28	20 s
#12. Singla et al. (2019b)	Sect. 3.2.8	VR 2019	ACR	4K, FHD	25–30	H.264	HTC Vive Pro	2880×1600	30	60 s

[a] Tests #4, #5, and #6 were published in a single paper
[b] Tests #8 and #9 were published in a single paper

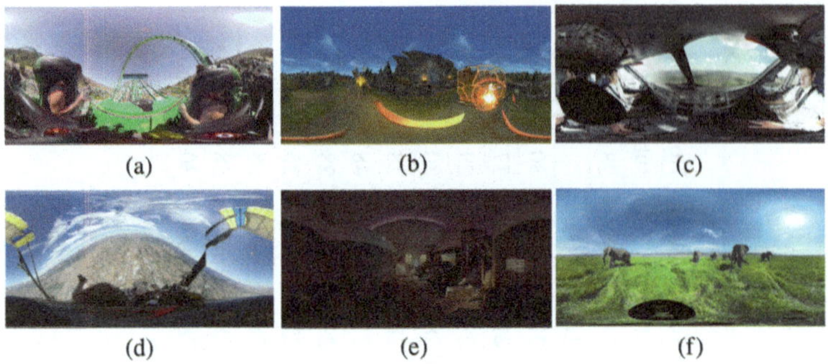

Fig. 3.2 Sample 2D representation scenes of 360° videos used in test 1. (**a**) Mega Coaster. (**b**) Project 360. (**c**) 360 Cockpit View. (**d**) Sky Diving. (**e**) Etihad A380. (**f**) Elephants

Table 3.2 Description of the source contents for Test 1 (Singla et al., 2017b)

Type	No.	Name	Resolution	FPS	Bitrate (in mbps)	Timestamp
Fast	1	Mega Coaster	3840 × 2048	30	17.7	00:10–01:10
			1920 × 1080	30	3.7	
	2	Project 360	3840 × 2160	30	17.4	01:36–02:40
			1920 × 1080	30	3.4	
Medium	3	360 Cockpit View	3840 × 1920	25	9.8	07:35–08:40
			1920 × 1080	25	3.6	
	4	Sky Diving in 360	3840 × 1920	29	16.4	03:22–04:25
			1920 × 1080	29	4.5	
Slow	5	Reimagine – Etihad A380	3840 × 2048	29	8.5	03:29–04:31
			1920 × 1080	29	2.4	
	6	Surrounded by Wild Elephants	3840 × 2048	30	13.0	01:19–02:24
			1920 × 1080	30	3.5	

encoding was applied. The inspection of the downloaded videos did not reveal any visible coding degradations. The downloaded dataset was subjectively categorized into three sets based on motion: slow, medium, and fast. Under the term "motion," both camera and content motion have been considered. Table 3.2 shows a content identifier, the content name, resolution, and frames per second. Each video sequence was cut with *XMedia Recode*, without changing the quality, to the target duration of 60–65 s.[1] The length was chosen to elicit possible symptoms of simulator sickness and make sure that the overall test duration did not exceed 90 min.

For the selected video sequences, Spatial (SI) and Temporal (TI) Information were analyzed based on P.910 (2008). Figure 3.3 shows the average (full dot) as well as per frame (circles) SI and TI values for the selected source video sequences.

[1] Please note that no further encoding was done after downloading the videos from YouTube.

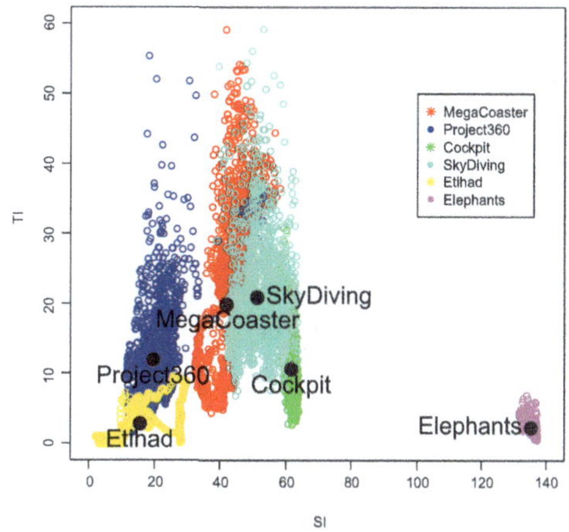

Fig. 3.3 SI vs. TI (P.910, 2008) of the tested videos for Test 1. The average (full dot) as well as per frame (circles) SI and TI values are shown to determine scene complexity and motion magnitude

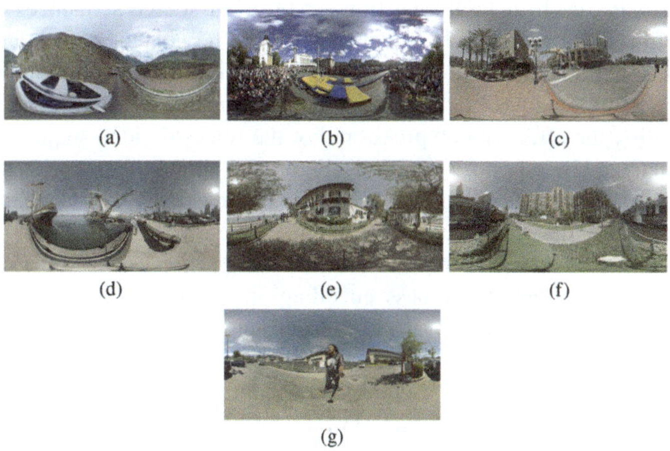

Fig. 3.4 Sample 2D representation scenes of 360° videos used in Tests 2–6. (**a**) DrivingInCountry. (**b**) PoleVault_le. (**c**) Gaslamp. (**d**) Harbor. (**e**) KiteFlite. (**f**) Trolley. (**g**) SkateboardInLot

3.2.2 Test 2 and Test 3

This section describes Tests 2 and 3. Tests 2 and 3 were published in ACM MM 2017 (Singla et al., 2017a) and HVEI 2018 (Singla et al., 2018), respectively. These two tests have a goal of subjective evaluation for different resolutions and bitrates using M-ACR and DSIS test methods under different contextual conditions.

For Tests 2 and 3, six different videos were selected without audio from Asbun et al. (2016), Schwarz et al. (2016), and Sun and Guo (2016), see Fig. 3.4. The NVIDIA GTX 980 graphics card used in the setup at the time can display videos up to 4096×2160@60. As a result, 8K sequences were downscaled to Ultra-High-Definition (UHD) and FHD resolution. The test video sequences are in YUV 4:2:0

Table 3.3 Source contents and IDs for Tests 2 and 3 (Singla et al., 2017a)

ID	Name
1	DrivingInCountry
2	PoleVault_le
3	Gaslamp
4	Harbor
5	KiteFlite
6	Trolley

Table 3.4 Target bitrates in MBit/s for different resolutions for Tests 2 and 3 (Singla et al., 2017a)

Resolution and bitrate	R1	R2	R3	R4	R5
UHD (3840×1920)	1	2	4	8	15
FHD (1920×1080)	0.5	1	1.5	3.5	5.0

color space, 8-bits per channel. Two categories of spatial resolution are addressed, UHD and FHD, each in ERP. The frame rate of all source sequences (SRCs) is 30 Hz, with a duration of each sequence of 10 s. The content names and content identifiers are shown in Table 3.3, and the target bitrates are shown in Table 3.4. The selection was mainly motivated by the requirements for source video quality when studying the effect of compression. For the selected video sequences, SI and TI were analyzed based on P.910 (2008). Figure 3.5 shows the average (full dot) as well as per frame (circles) SI and TI values for the selected video sequences.

For encoding the sequences with High-Efficiency Video Coding (HEVC), `FFmpeg` was used with *libx265* and the *Video Buffering Verifier* (VBV) method. It uses bitrate-constrained one-pass encoding and ensures that the target bitrate is constrained to a maximum value by specifying both a maximum bitrate and client buffer size. This encoding scheme reflects a basic encoding setting useful for streaming; it is considered a good starting point for the test series conducted in the Joint Video Exploration Team (JVET) during standardization at the time of tests (Boyce, 2017). `libx265` provides advantages over the `HM`[2] reference software in terms of high encoding speed and advanced rate control. For each content and resolution, five different target bitrates were selected, corresponding to different quality levels (see Table 3.4).

3.2.3 *Test 4, Test 5, and Test 6*

This section describes Tests 4, 5, and 6. These tests were published in MMSP 2019 (Singla et al., 2019a). The goal of these three tests was to compare the performance of three different test methods, ACR, M–ACR, and DSIS based on the statistical reliability and assessment time under the same contextual conditions.

[2] https://hevc.hhi.fraunhofer.de/svn/svn_HEVCSoftware/ (Accessed on 2017/03/15).

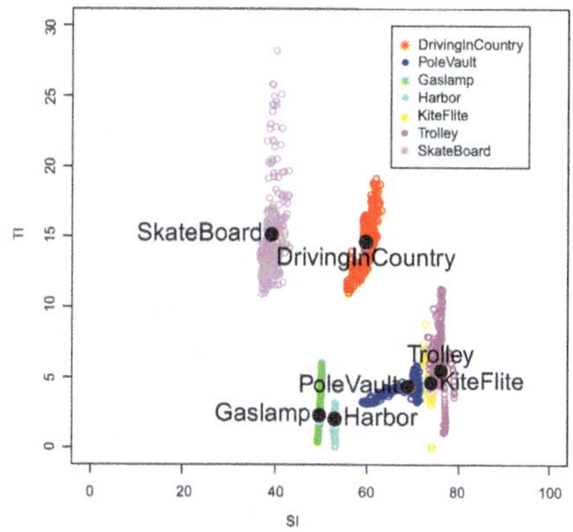

Fig. 3.5 SI vs. TI of the tested videos for Tests 2–6. The average (full dot) as well as per frame (circles) SI and TI values are shown to determine scene complexity and motion magnitude

Table 3.5 Target bitrates in MBit/s for different resolutions for Tests 4, 5, and 6 (Singla et al., 2019a)

Resolution and bitrate	R1	R2	R3	R4
4K (3840×1920)	0.5	1	4	8
6K (6144×3072)	2	4	10	15
8K (7680×3840)	6	10	18	25

Five different video sequences (1:`Gaslamp`, 2:`Harbor`, 3:`KiteFlite`, 4:`Skateboard_in_lot`, 5:`Trolley`) were selected without audio from the standard 360° video dataset, published by the JVET (Abbas and Adsumilli, 2016; Asbun et al., 2016), see Fig. 3.4. The resolution and frame rate of the video sequences are 8192×4096 and 30 fps, respectively, with 10 s duration. All the test sequences are in YUV 420 color space, 8-bits per channel, and ERP format. The NVIDIA GTX 1080 TI graphics card used in the tests at the time can display videos up to 7680×4320@60. Therefore, all the test video sequences were down-sampled to 7680×3840 (Lopes et al., 2018). Also, the employed HMD (HTC Vive Pro) cannot play 10-bit video, so the 10-bit video sequence #4 was converted into 8-bit using FFmpeg. The information about the target bitrates for each resolution is given in Table 3.5. Different sets of bitrate–resolution pairs were selected to achieve a balanced distribution of the users' ratings over the entire range of the 5-point scale of ACR, DSIS, and M-ACR test methods. For the selected video sequences, SI and TI were analyzed based on P.910 (2008). Figure 3.5 shows the average (full dot) as well as per frame (circles) SI and TI values for the selected video sequences.

For encoding the source videos, `FFmpeg`[3] with *libx265* (HEVC) was applied. Two-pass (preset slow) encoding with the specified target bitrate was used.

[3] https://ffmpeg.org/ (Accessed on 2017/03/15).

3.2.4 Test 7

Test 7 has primarily been conducted by the first author (Stephan Fremerey) and published in ACM MMSys 2018 (Fremerey et al., 2018). It is included in this book to provide a complete set of information on all tests the book author has contributed to. The goal of this test was to create a head rotation dataset of subjects watching different omnidirectional videos in a task-free scenario. Additionally, AVTrack360 was published to conduct subjective tests and record head rotation data.

For the source material of Test 7, one important aspect was to choose a variety of SRCs. The subjects should stay entertained during the test and keep watching the videos in an exploratory manner. The SRCs were used in 4K resolution taken from YouTube and ARTE, while 6 are downloaded from YouTube and 14 from ARTE, see Fig. 3.6. All source videos were acquired in ERP format, YUV 420 color space, and 8-bits per channel. In Table 3.6, the respective timestamps for starting cutting a 30 s extract are provided. The links to the SRCs can be found in the Readme file of the dataset.[4] youtube-dl[5] was used to download the contents from YouTube. In addition, to not influence the behavior by the integral video quality, it was important that all SRCs roughly have the same visual quality. For cutting the 30 s test sequences and inserting a 10 s gray screen afterward, FFmpeg was used. The SRCs were converted into raw video, concatenated with a 10 s gray sequence and afterward converted back to *libx264* (H.264) lossless using the *-crf 0* parameter, so no additional, perceivable quality degradation would be introduced. The audio was removed for all test sequences to have no additional influence by this factor.

For the selected video sequences, SI and TI were analyzed based on P.910 (2008). Figure 3.7 shows the average (full dot) as well as per frame (circles) SI and TI values for the selected video sequences.

3.2.5 Test 8 and Test 9

This section describes Tests 8 and 9. These tests were published in ACM MMSys 2019 (Singla et al., 2019c). The goal of these tests was to present a testbed and describe an end-to-end test method for the quality evaluation of viewport-adaptive streaming of 360° videos. Furthermore, the subjective quality and simulator sickness was assessed and compared for two different scenarios: (1) resolution- and bandwidth-limited, (2) bandwidth-limited only, for different simulated network latency values.

Three different omnidirectional videos (without audio) were used, see Fig. 3.8. Each video was provided by a different content provider. Table 3.7 lists information

[4] https://github.com/acmmmsys/2018-AVTrack360

[5] https://rg3.github.io/youtube-dl (Accessed on 2018/01/10).

Fig. 3.6 Sample 2D representation scenes of 360° videos used in Test 7. (**a**) Icebreaker. (**b**) Fireworks. (**c**) Helicopter. (**d**) Hawaii. (**e**) Penguins. (**f**) Diving. (**g**) Colonie. (**h**) Childs. (**i**) Wasala. (**j**) Carrara. (**k**) EdgeofSpace. (**l**) Arcachon. (**m**) LaGrotteChauvet. (**n**) Aliens. (**o**) Elisir. (**p**) SpotlightHelp. (**q**) AngelFalls. (**r**) Holifest. (**s**) NerfBattle. (**t**) Football

about these video sequences such as resolution, frame rate, and the name of the sequences along with their content identifier. All the test sequences are in YUV 420 color space, 8-bits per channel, and ERP format. The duration of each video sequence was chosen to be 30 s, so that subjects have enough time to explore the video because, in tile-based streaming, the response to head rotation plays an important role in the overall QoE.

For the selected video sequences, SI and TI were analyzed based on P.910 (2008) to ensure that the selected video sequences had different SI and TI characteristics across all scenes. Figure 3.9 shows the average (full dot) as well as per frame (circles) SI and TI values for the selected video sequences in their original resolutions (see Table 3.7).

For encoding the anchor videos in full resolution using the HEVC encoder, `FFmpeg`[6] was used with *libx265* (HEVC) and the *Video Buffering Verifier* (VBV) method. Due to hardware limitations, the videos were playable in full resolution

[6] https://ffmpeg.org/ (Accessed on: 2018/06/01).

Table 3.6 Source contents used in Test 7 (Fremerey et al., 2018)

ID	Name	Timestamp
1	Icebreaker	00:01:03
2	Fireworks	00:01:40
3	Helicopter	00:04:01
4	Hawaii	00:01:52
5	Penguins	00:01:39
6	Diving	00:02:16
7	Colonie	00:01:55
8	Childs	00:01:18
9	Wasala	00:05:47
10	Carrara	00:02:07
11	Edge of Space	00:00:10
12	Arcachon	00:03:05
13	La Grotte Chauvet	00:03:10
14	Aliens	00:05:47
15	Elisir	00:01:50 (18.7s) 00:02:25 (11.3s)
16	Spotlight HELP	00:01:10
17	Angel Falls	00:00:12
18	Holifest	00:06:17
19	Nerf battle	00:00:13
20	Football	00:00:16

up to 6K. Different anchor conditions were used, including resolutions and bitrates. Information about the anchor conditions corresponding to the different tests (8 and 9) are shown in Table 3.8. The bitrate for 4K resolution was chosen based on the experiment conducted in Singla et al. (2017a, 2018) of Tests 2 and 3. For 6K resolution, 20 Mbit/s was chosen based on Tests 4–6 (Singla et al., 2019a), whereas, for Test 9, a much higher bitrate 40 Mbit/s was chosen. The anchor conditions were compared with conditions of tile-based streaming, as discussed in more detail in Sect. 5.3.

3.2.6 *Test 10*

This section describes Test 10. This test was published in IEEE VR 2020 (Raake et al., 2020). The goal of this test was to present the first general concept of the simulator sickness model applicable to complete viewing sessions, and prediction of simulator sickness over time, taking breaks into account.

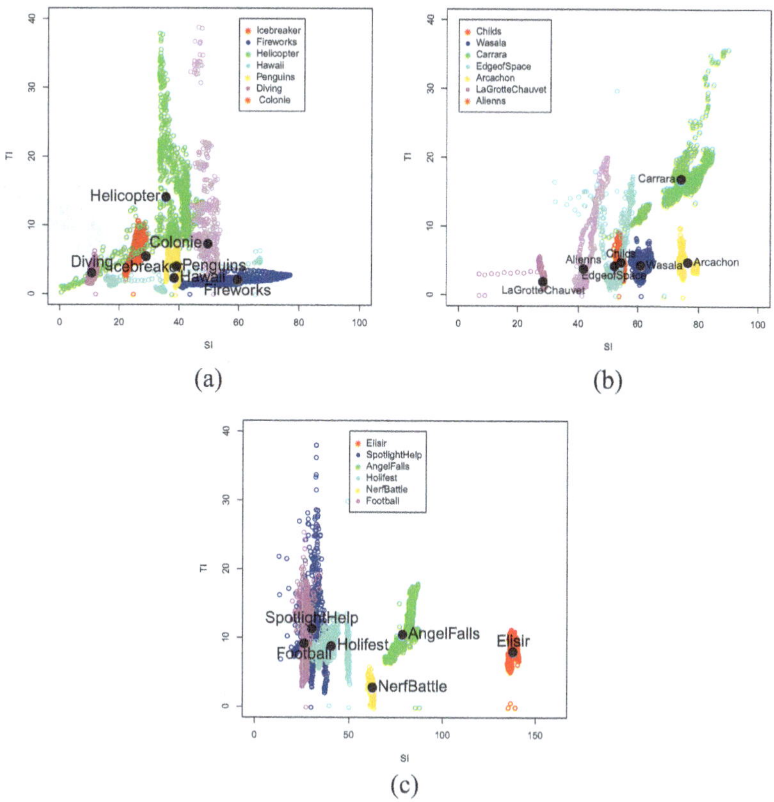

Fig. 3.7 SI vs. TI of the tested videos for Test 7. The average (full dot) as well as per frame (circles) SI and TI values are shown to determine scene complexity and motion magnitude. (**a**) Contents 1–7. (**b**) Contents 8–14. (**c**) Contents 15–20

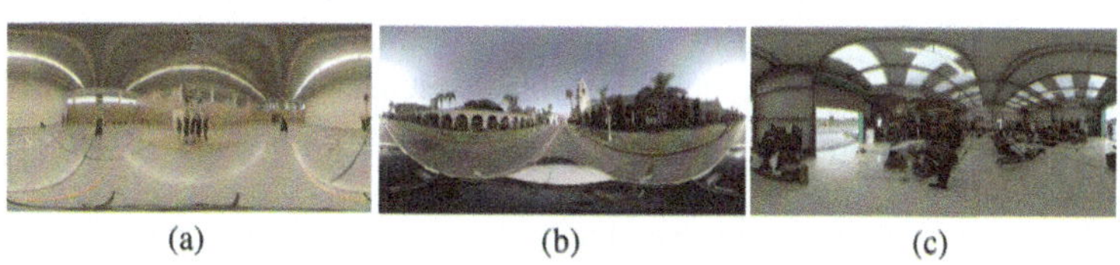

Fig. 3.8 Sample 2D representation scenes of 360° videos used in Tests 8 and 9. (**a**) CheerLeading. (**b**) Balboa. (**c**) F3

Table 3.7 Information of the test sequences for Tests 8 and 9 (Singla et al., 2019c)

ID	Name	Resolution	Frame rate (frames per second)
1	CheerLeading	7680 × 3840	25
2	Balboa	6144 × 3072	60
3	F3	8192 × 4096	25

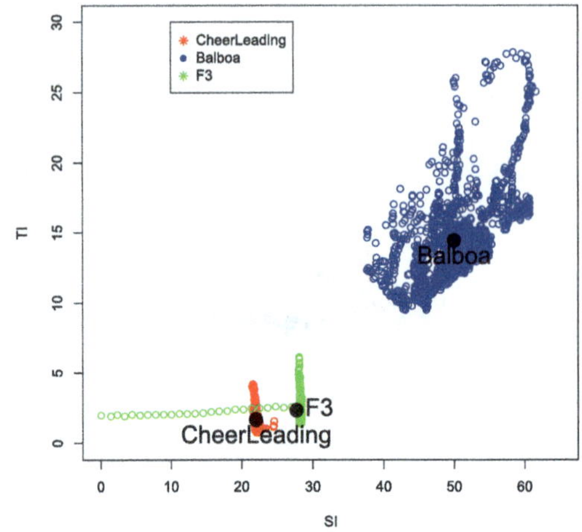

Fig. 3.9 SI vs. TI of videos for Tests 8 and 9 (Singla et al., 2019c). The average (full dot) as well as per frame (circles) SI and TI values are shown to determine scene complexity and motion magnitude

Table 3.8 Target bitrates for different anchor conditions for Tests 8 and 9 (Singla et al., 2019c)

	Test 8		Test 9
Resolution	4K	6K	6K
Bitrate (Mbit/s)	8	20	40

Eighteen different omnidirectional source videos with audio were used in the test,[7] see Fig. 3.10. The video sequences were downloaded from YouTube/Arte in 4K (3840 × 1920) resolution in the highest available quality. All the sequences are in YUV 420 color space, 8-bits per channel, and ERP format. The duration of each video sequence is 60 s, to provide sufficient time for the subjects to explore the scene and possibly develop symptoms of simulator sickness. The videos were re-encoded with HEVC using `FFmpeg` with *libx265* and two-pass (preset slow) encoding in two bitrates (1 and 15 Mbit/s). These bitrates were chosen to investigate the impact of bitrates on comfort level as a measure of the induced simulator sickness. The source videos were selected to include some sequences expected to make the user uncomfortable, e.g., due to high amounts of motion as a measure for the induced simulator sickness.

For the selected video sequences, SI and TI were analyzed based on P.910 (2008). Figure 3.11 shows the average (full dot) as well as per frame (circles) SI and TI values for the selected video sequences.

[7] https://zenodo.org/record/3661134#.XkFxrDFKiUk

Fig. 3.10 Sample 2D representation scenes of 360° videos used in Test 10. (**a**) Antartica1. (**b**) Antartica2. (**c**) Antartica3. (**d**) Arcachon. (**e**) Wolfskinder. (**f**) Carrara. (**g**) RevoCollection. (**h**) Elisir. (**i**) VRWorks. (**j**) Project360. (**k**) MegaCoaster. (**l**) MessaDa. (**m**) MontBlanc. (**n**) Fontainebleau. (**o**) Splash16. (**p**) Teleportaled. (**q**) Elephants. (**r**) WinterVenice

3.2.7 Test 11

This section describes Test 11. This test was published in IEEE TMM 2021 (Gutierrez et al., 2022). In Test 11, 10 subjective tests were conducted by ten research labs. It is included in this book to provide information about the test that was conducted by the author of the book. The goal of this test was to compare two test methods (ACR vs. DSIS) with different durations of PVSs on perceived video quality. Furthermore, the influence of HMDs, audio and scoring methods on the visual quality of 360° videos was also investigated.

Eight 360° videos of 20 s were used in the test, see Fig. 3.12. They were all in ERP, monoscopic, and had at least a resolution of 3840 × 1920 pixels and 25 fps. Table 3.9 lists the properties of the source sequences used in Test 11. The original videos were provided by Nokia, TU Ilmenau (see Sect. 3.3), VSense (Knorr et al., 2018), and Oculus. The selected sequences present a wide range of content characteristics, including one video with camera motion video referred to as

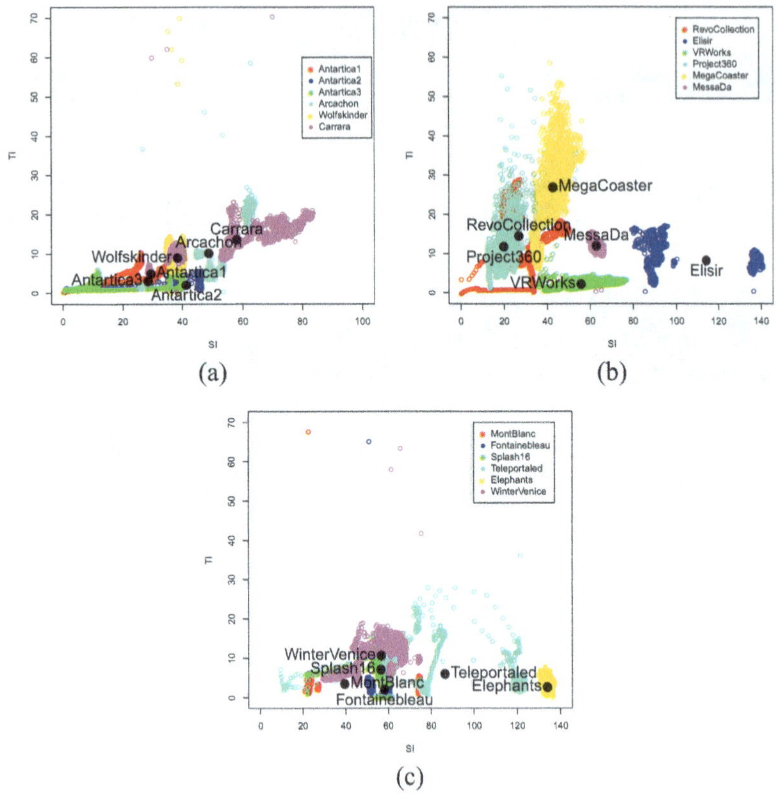

Fig. 3.11 SI vs. TI of the tested videos for Test 10. The average (full dot) as well as per frame (circles) SI and TI values are shown to determine scene complexity and motion magnitude. (**a**) Contents 1–6. (**b**) Contents 7–12. (**c**) Contents 13–18

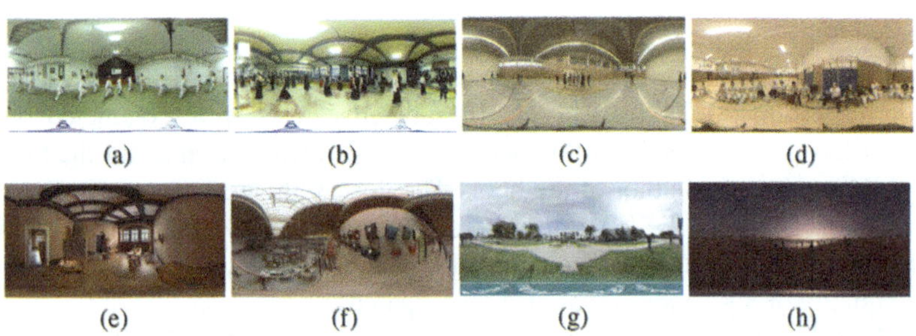

Fig. 3.12 Sample 2D representation scenes of 360° videos used in Test 11 (Gutierrez et al., 2022). (**a**) NokiaDojo. (**b**) NokiaFlamenco. (**c**) CheerLeading. (**d**) BrazilMusic. (**e**) VSenseLuther. (**f**) VSenseVaude. (**g**) OculusMotion. (**h**) OculusBeach

"OculusMotion," one with animation scenes (VSenseLuther), and different spatial and temporal complexities, as shown in Fig. 3.13.

Table 3.9 Properties of the source sequences used in Test 11

Seq_Id	Content Name	Resolution	Frame rate	Provider
1	NokiaDojo (ND)	3840 × 2160	30 fps	Nokia
2	NokiaFlamenco (NF)	3840 × 2160	30 fps	Nokia
3	CheerLeading (CL)	4096 × 2048	25 fps	TU Ilmenau
4	BrazilMusic (BM)	4096 × 2048	25 fps	TU Ilmenau
5	VSenseLuther (VL)	4096 × 2048	30 fps	VSense
6	VSenseVaude (VV)	4096 × 2048	30 fps	VSense
7	OculusMotion (OM)	3840 × 2160	30 fps	Oculus
8	OculusBeach (OB)	3840 × 2160	30 fps	Oculus

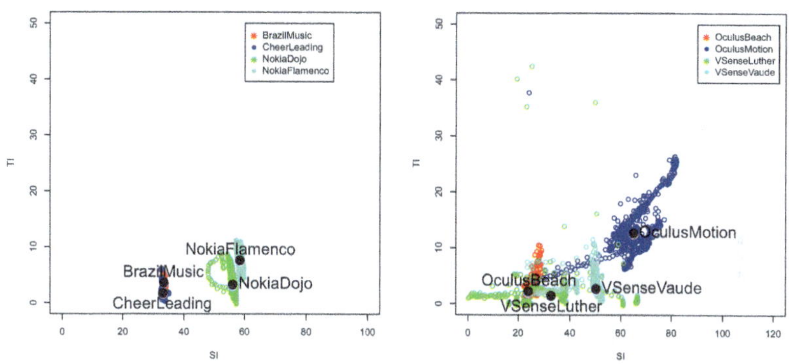

(a) (b)

Fig. 3.13 SI vs. TI of the tested videos for Test 11. The average (full dot) as well as per frame (circles) SI and TI values are shown to determine scene complexity and motion magnitude. (**a**) Contents 1–4. (**b**) Contents 5–8

#Tiles	Transition	ROI	QPs					
6x3	Smooth	120°	42	32	22	22	32	42
6x3	Abrupt	120°	37	37	22	22	37	37

Fig. 3.14 Settings for the non-uniform coding configurations for Test 11 (Gutierrez et al., 2022)

Eight different HEVC coding configurations were applied to generate the test videos, including four uniform encodings (using homogeneous QPs) and four non-uniform encodings (using different configurations of tiles). For the uniform configurations, the following QPs were used: 15, 22, 32, 42, while Fig. 3.14 shows the settings for the non-uniform ones. It can be seen, two different structures of tiles were used, and smooth and abrupt transitions between adjacent tiles were considered. The encoding of all the test sequences was done using the Kvazaar encoder. Please note that there was no tile-based streaming involved here. Only different QPs were used to encode different regions of video.

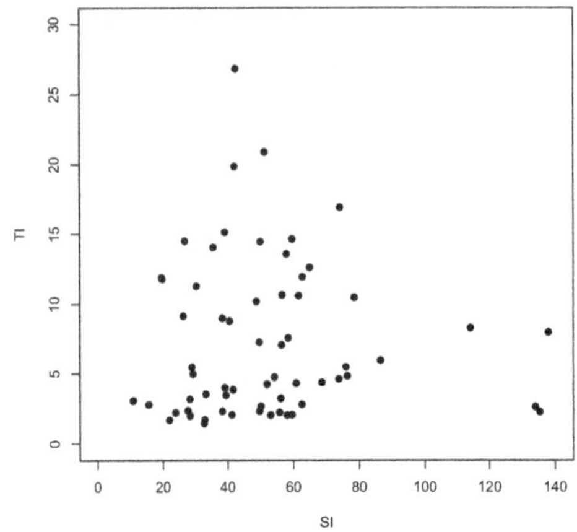

Fig. 3.15 The average values of SI vs. TI of the tested videos for Tests 1–12

3.2.8 *Test 12*

For Test 12, the Test 1 data were used. Please refer to Sect. 3.2.1 for more information. A series of tests was presented comparing different test protocols to assess integral quality, simulator sickness, and presence for omnidirectional videos. In addition, the well-established Simulator Sickness Questionnaire and Presence Questionnaire methods were used, once in a full version, and once with only one single integral scale, to analyze how well presence and simulator sickness can be captured using only a single scale.

3.2.9 *Summary of Used 360° Source Videos*

Figure 3.15 shows the scatter plot of average SI compared to TI values for the videos used in Tests 1–12. It can be seen from Fig. 3.15 that most of the sequences have lower temporal complexity and lower to medium spatial complexity. Only a few of the sequences have high temporal complexity with values above 15 and have medium spatial complexity. A very few sequences have high spatial complexity with values more than 100; furthermore, they have lower temporal complexity.

3.3 Generation of 360° Videos

Recently, several new video coding algorithms have been proposed more suitable for high resolution, e.g., UHD–1/2, (e.g., AV1, HEVC, or Versatile Video Coding (VVC)). To evaluate such high-quality codecs, also high-quality source content is required. However, one essential challenge is that during the time of working on

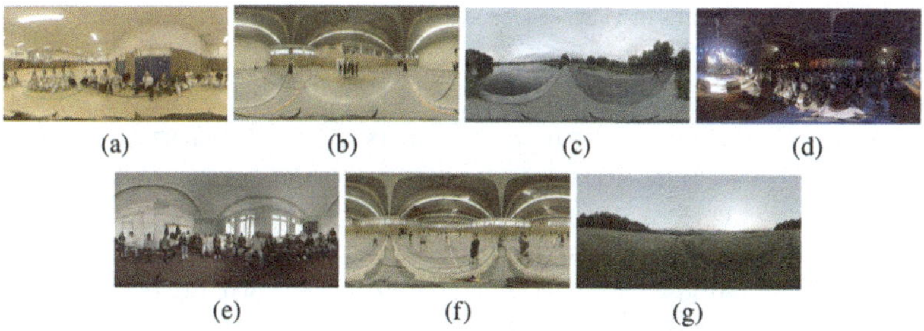

Fig. 3.16 Sample 2D representation scenes of 360° video proposed in the book. (a) BrazilMusic. (b) CheerLeading. (c) ThievesAtLake. (d) Concert. (e) GroupSong. (f) Sports. (g) Sunset

this book, there were rather few high-quality publicly available source contents (see Sect. 2.2). In order to evaluate and compare different encoding algorithms, there is a need for high-resolution and longer-duration source contents not only for 2D video, but also for 360° video as studied in this book. A new set of high-quality 360° video sequences captured using a GoPro Omni camera is proposed in this book. The proposed dataset contains seven 360° videos captured in 8K (7680×3840) resolution each with a duration of 30 s, with no scene cuts, see Fig. 3.16. Besides the acquired videos, various features such as spatial and temporal information (SI and TI), contrast (Göring et al., 2019), FFT (Göring et al., 2019; Katsavounidis et al., 2015), and NIQE (Mittal et al., 2013) associated with each video sequence were calculated.

The database contains seven videos recorded with a stationary GoPro Omni camera. The GoPro Omni rig has an array of six synchronized GoPro Hero4 Black cameras. To stitch the videos, Kolor Autopano Video Pro was used. After stitching, no encoding was performed on the videos. The resulting stitched videos are provided in ERP format in an AVI container. The videos were exported from the cameras with a total resolution of 7680×3840, a frame rate of 25 fps with audio.[8] This database includes scenes such as sports, cheerleading, group singing, a concert, or Brazilian music, where a viewer is expected to explore the scene in the whole horizontal direction (360°), and for some sequences such as cheerleading, the viewer is expected to also move their head vertically. For some video sequences, such as thieves at the lake, the viewer will try to follow the story and will look in certain directions only. All of the video sequences may not elicit any symptoms of simulator sickness in viewers, as the motion magnitude (temporal complexity) is quite low in the proposed videos.[9]

Table 3.10 summarizes the features that are calculated for each video individually. At first, SI/TI (P.910, 2008) was extracted. For all videos, the SI values are

[8] Audio was recorded using the in-built microphone of the cameras.

[9] Please note that this is also a drawback of the proposed videos. If in a subjective test, some sequences are expected to make users uncomfortable. These videos are not appropriate.

Table 3.10 Different features of video sequences

Video	SI	TI	Contrast	FFT	NIQE
BrazilMusic	26.2656	4.04224	83.74	291388.64	20.44
CheerLeading	26.3936	2.0736	85.15	298300.876	20.33
ThievesAtLake	37.9136	1.344	85.2	301538.744	19.59
Concert	32.1536	3.2	84.195	301998.265	20.41
GroupSong	27.52	3.34592	89.88	298982.407	20.33
Sports	30.464	7.3472	92.98	300212.218	20.23
Sunset	27.9552	0.9728	91.29	303145.99	20.07

within a similar range [26, 38], and the TI values in a range [0, 8], because the introduced videos include less motion.[10] Moreover, a contrast feature, based on contrast histogram equalization (Göring et al., 2019), is calculated. High values refer to a good contrast of the videos and can be confirmed by all included videos. Furthermore, an FFT feature (Göring et al., 2019; Katsavounidis et al., 2015) was calculated, which is a native resolution detector, where higher values refer to a higher native resolution. As the last feature, the NIQE (Mittal et al., 2013) was selected, as a no-reference video quality feature, where for the proposed videos the values are within the same range, indicating a high source video quality. The self-produced videos are available at https://avtshare01.rz.tu-ilmenau.de/asingla/PhD360Video/.

3.4 Test Environment

To avoid the influence of external factors such as environmental noise, a controlled lab environment was set up to conduct the subjective tests. Before starting a test, all the curtains in the room were closed so that there should not be any sunlight present in the test room, as it may interfere with an HMD tracking system. The base stations were kept in opposite corners, and their height was set to 2 m, in order to get an unobstructed view of each other. When a viewer wears the HMD, IPD (Interpupillary Distance) or focus and straps of the HMD were adjusted according to the viewer. The viewer was always seated on a swivel chair to freely explore the 360° videos. The test supervisor (or experimenter) was always present in the room to note down the video quality ratings and to pause or stop the subjective test in case of the viewer experiences the symptoms of simulator sickness. Please note that subjects were given instructions not to consider the ghost and stitching artifacts while rating the video quality. An example image (in color) of ghost and stitching

[10] If in a video quality test, higher motion videos are sought, then these videos would not be suitable.

artifacts was taken as a snapshot from a video printed on an A4 sheet was shown to the participants.

3.5 AVTRack360

AVTrack360 was developed by Fremerey et al. in (2018) and is publicly available.[11] It is a Python command-line tool for recording head rotation values of each axis, Pitch/Yaw/Roll, respectively, for HTC Vive (Pro), and Oculus Rift CV1. It runs on the Windows 7+ Operating System (OS) and records the orientation angles every 5 ms in synchronization with the video playback. As AVTrack360 is written in Python, it can easily be set up by installing the respective packages or a virtual Python environment as explained in the Readme file. In Fig. 3.17, the general concept of AVTrack360 is shown. After loading the respective playlist file in JavaScript Object Notation (JSON) format and gathering the required duration of the video using FFmpeg, the 360° video is played back. To this aim, the Whirligig player[12] (Day, 2018) was used in all of the tests except in Tests 8, 9, and 11. In Tests 8 and 9, ClearVR player is used, and for Test 11, Miro360 (Pérez and Escobar, 2019) was used for the playout of the 360° videos.

For enabling smooth video playout and supporting the used codec, Whirligig was configured to use the Open Source LAV DirectShow media decoder[13] v0.70.2. Whirligig was chosen because of its good audiovisual playback quality, performance, and command-line functionality. Hence, AVTrack360 can start the playback correctly by setting the respective playout parameters specified in the playlist, such as the input file or projection scheme to use. Because Whirligig cannot receive any commands via an API, after the player has loaded the video, starting the playback is managed by AVTrack360 simulating a keystroke. To assure every viewer is starting to watch the video at exactly the same position, yaw = 0, and the viewport gets reset to the initial position directly before the playout by simulating a keystroke. AVTrack360 gets the current tracking positions of the HMD accessing

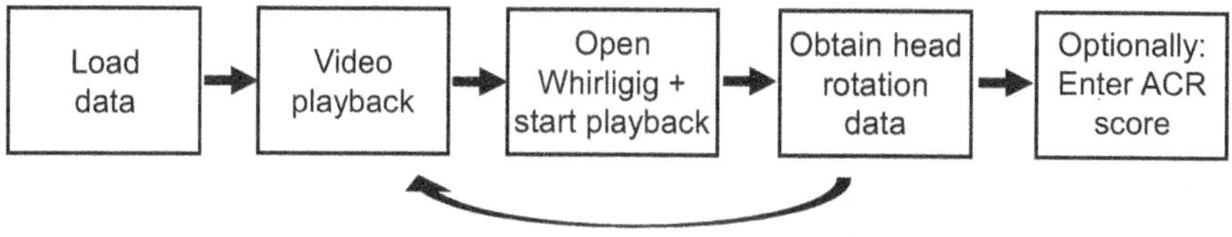

Fig. 3.17 Concept of AVTrack360 (Fremerey et al., 2018)

[11] https://github.com/acmmmsys/2018-AVTrack360

[12] The FOV and projection type were set to *Barrel* and *360*, respectively, in the *Whirligig* player.

[13] https://github.com/Nevcairiel/LAVFilters

the functionality of the Open Source Python bindings of the respective Software Development Kits (SDKs). For Oculus Rift pyovr,[14] for HTC Vive pyopenvr[15] is used. The playback time of the video is saved along with the current head rotation data in a JSON file. When the measured time exceeds the video duration, the recording gets stopped and Whirligig is exited at the same time. If enabled via the respective parameter, AVTrack360 asks the test supervisor to enter an ACR (P.910, 2008) score for the video shown to the participant. Afterward, the next video of the playlist is loaded and the described procedure is repeated.

For avoiding presentation-order effects, randomized and dedicated playlists were used for each test participant for all the tests.

3.6 Conclusions

The dataset used in this book is partially publicly available, e.g., all the videos that were downloaded from YouTube/Arte can be accessed and downloaded easily from YouTube/Arte. The videos that were downloaded from JVET have restricted access and are password protected.

The self-produced videos can be downloaded using this link: https://avtshare01.rz.tu-ilmenau.de/asingla/PhD360Video/. All the test-related data such as video quality, simulator sickness, and head rotation can be accessed from the: https://avtshare01.rz.tu-ilmenau.de/asingla/PhDData/.

The software (AVTrack360) is used to conduct the subjective test and to track the head rotation data that are publicly available on GitHub: https://github.com/acmmmsys/2018-AVTrack360.

References

A. Abbas, B. Adsumilli, *AHG8: New GoPro Test Sequences for Virtual Reality Video Coding*. JVET-D0026, 4th Meeting, Chengdu, 15–21 Oct 2016

E. Asbun et al., *AHG8: InterDigital Test Sequences for Virtual Reality Video Coding*. JVET-D0039, 4th Meeting, Chengdu, 15–21 Oct 2016

J. Boyce, AhG8: update on JCT-VC and JVET 360 video activities, in *Joint Video Exploration Team of ITU-T SG16 WP3 and ISO/IEC JTC1/SC29/WG11, JVET-E0137*, 5th Meeting, Oct 2017

P. Day, *Whirligig Free – Whirligig*, 2018. http://www.whirligig.xyz/new-page-3

S. Fremerey et al., AVtrack360: an open dataset and software recording people's head rotations watching 360-degree; videos on an HMD, in *Proceedings of the 9th ACM MMSys*, Amsterdam, 2018. ISBN: 978-1-4503-5192-8. https://doi.org/10.1145/3204949.3208134

[14] https://github.com/cmbruns/pyovr

[15] https://github.com/cmbruns/pyopenvr

S. Göring et al., Analyze and predict the perceptibility of UHD video contents, 2019. https://doi.org/10.2352/ISSN.2470-1173.2019.12.HVEI-215

J. Gutierrez et al., Subjective evaluation of visual quality and simulator sickness of short 360° videos: ITU-T Rec. P.919. IEEE Trans. Multimedia **24**, 3087–3100 (2022). https://doi.org/10.1109/TMM.2021.3093717

I. Katsavounidis, A. Aaron, D. Ronca, Native resolution detection of video sequences, in *Annual Technical Conference and Exhibition, SMPTE 2015*, pp. 1–20

S. Knorr et al., Director's cut – a combined dataset for visual attention analysis in cinematic VR content, in *European Conference on Visual Media Production*, Dec 2018

F. Lopes et al., Subjective and objective quality assessment of omnidirectional video, in *SPIE Conference Series*, Sept 2018. https://doi.org/10.1117/12.2321679

A. Mittal, R. Soundararajan, A.C. Bovik, Making a "completely blind" image quality analyzer. IEEE Signal Process. Lett. **20**(3), 209–212 (2013)

ITU-T Rec. P.910, *Recommendation ITU-T P.910: Subjective Video Quality Assessment Methods for Multimedia Applications*, 2008

P. Pérez, J. Escobar, MIRO360: a tool for subjective assessment of 360 degree video for ITU-T P.360-VR, in *2019 Eleventh International Conference on Quality of Multimedia Experience (QoMEX)*, June 2019, pp. 1–3. https://doi.org/10.1109/QoMEX.2019.8743216

A. Raake et al., SiSiMo: towards simulator sickness modeling for 360-degree videos viewed with an HMD, in *2020 IEEE Conference on Virtual Reality and 3D User Interfaces Abstracts and Workshops (VRW)*, Mar 2020, pp. 582–583

S. Schwarz, A. Aminlou et al., *Tampere Pole Vaulting Sequence for Virtual Reality Video Coding*. JVET-D0143, 4th Meeting, Chengdu, 15–21 Oct 2016

A. Singla et al., Comparison of subjective quality evaluation for HEVC encoded omnidirectional videos at different bit-rates for UHD and FHD resolution, in *Proceedings of the on Thematic Workshops of ACM Multimedia*, Mountain View, 2017a. ISBN: 978-1-4503-5416-5. https://doi.org/10.1145/3126686.3126768

A. Singla et al., Measuring and comparing QoE and simulator sickness of omnidirectional videos in different head mounted displays, in *9th International Conference on QoMEX*, May 2017b, pp. 1–6. https://doi.org/10.1109/QoMEX.2017.7965658

A. Singla, W. Robitza, A. Raake, Comparison of subjective quality evaluation methods for omnidirectional videos with DSIS and Modified ACR. Electron. Imaging **2018**, 1–6 (2018). https://doi.org/10.2352/ISSN.2470-1173.2018.14.HVEI-525

A. Singla, W. Robitza, A. Raake, Comparison of subjective quality test methods for omnidirectional video quality evaluation, in *2019 IEEE 21st International Workshop on Multimedia Signal Processing (MMSP)*, Sept 2019a, pp. 1–6

A. Singla et al., Assessing media QoE, simulator sickness and presence for omnidirectional videos with different test protocols, in *2019 IEEE Conference on Virtual Reality and 3D User Interfaces (VR)*, Mar 2019b, pp. 1163–1164. https://doi.org/10.1109/VR.2019.8798291

A. Singla et al., Subjective quality evaluation of tile-based streaming for omnidirectional videos, in *Proceedings of the 10th ACM Multimedia Systems Conference*, MMSys'19, Amherst (ACM, 2019c), pp. 232–242. ISBN: 978-1-4503-6297-9. https://doi.org/10.1145/3304109.3306218

W. Sun, R. Guo, *Test Sequences for Virtual Reality Video Coding from LetinVR*. JVET-D0179, 4th Meeting, Chengdu, 15–21 Oct 2016

Chapter 4
Comparison of Different Subjective Test Methods

4.1 Test Design

Considering HMD-viewing side effects such as simulator sickness and visual fatigue, the subjective test was split into four test sessions, where the duration of each session was no longer than 15 min. In each session, subjects rated fifteen 360° videos, and after every test session, subjects were asked to fill in the SSQ (Kennedy et al., 1993).

Two subjects performed the test one after the other. For hygiene reasons, the HMDs and their lenses were cleaned after every test session. The HMD name was also hidden to decrease contextual effects. Subjects were asked to watch the omnidirectional videos while sitting on a rotating chair so that they could explore the videos in any direction (3 DoF). AVTrack360 (Fremerey et al., 2018) is used in both experiments to capture the head rotation data of the test subjects.

Written instructions were given to the test participants explaining the test-related tasks. The test was divided into two parts: the training and test phase. In the training phase, the subjects were familiarized with the assessment procedure and the HMD was adjusted according to their head size. The test phases are described in Sects. 4.1.1 and 4.1.2 for experiments 1 and 2, respectively. For avoiding presentation-order effects, randomized and dedicated playlists were used for each test participant.

4.1.1 Experiment 1

Each subject had to rate 60 PVSs (six contents, two resolutions, and five bitrates). Each content was viewed 20 times (five bitrates, two resolutions, one repetition). In DSIS test method, the reference sequence and the PVS are compared, and in the

A. Singla, *Assessment of Visual Quality and Simulator Sickness for Omnidirectional Videos*, T-Labs Series in Telecommunication Services,
https://doi.org/10.1007/978-3-031-49988-3_4

M-ACR test method, each PVS is repeated twice (see the Sect. 4.3 for more details). Before the start of each test stimulus, the subjects were re-positioned in the center in order to start the recording of head motion data (yaw and pitch) from zero. The Oculus Rift HMD (Consumer Version 1) was used to watch the omnidirectional videos. The resolution and field of view of this device are 2160 × 1200 and 110°, respectively.

The total duration of the test for the M–ACR and DSIS was 120 and 100 min, respectively.

4.1.2 *Experiment 2*

Each subject had to rate 60 PVSs (five video sequences, three resolutions, and four bitrates). Each content was viewed 24 times (four bitrates, three resolutions, one repetition) except for the ACR test method, where each content was viewed 12 times.[1] In DSIS test method, the reference sequence and the PVS are compared, and in the M-ACR test method, each PVS is repeated twice (see the Sect. 4.3 for more details). After the completion of each test session, subjects were asked to fill in the SSQ (Kennedy et al., 1993) twice – once after the test session, and once again when the break was about to end. This was done to investigate the impact on how taking breaks help in reducing simulator sickness. The HTC Vive Pro was used to watch the omnidirectional videos. The resolution and field of view of this device are 2880 × 1600 and 110°, respectively.

The total duration of the test for the ACR, M–ACR, and DSIS was 70, 100, and 100 min, respectively.

4.2 Test Subjects

For all the tests, the test participants are recruited from the student body of the university. For each test, all participants were paid according to the duration of the tests. So, for a test with a duration of 60 min, participants were paid 10 Euros.

4.2.1 *Experiment 1*

30 subjects (15 female and 15 male) participated in the M–ACR viewing test. The subjects' age was between 19 to 36 years, with an average of 25.6 and a median of 25 years. 27 subjects (11 females and 16 males) participated in the DSIS subjective

[1] In the ACR test method, a PVS is watched only once.

test. The age of the participants varied from 19 to 36 years, with an average age of 25.1 and a median of 24 years.

4.2.2 Experiment 2

For the ACR test method, 29 subjects (18 male and 11 female) participated in the study. The participants' age was between 21 and 35, with an average of 26 years. For the M–ACR study, 30 subjects (16 male and 14 female) took part, aged between 19 and 33, with an average of 25.6 years. For DSIS, 28 subjects (19 male and 9 female) were recruited; their age was between 18 and 30, with an average of 25 years. There was no overlap between the subjects participating in these three test runs, according to a between-subject design chosen for comparing the test methods. This would help in avoiding the transfer of learning across different test methods as this experiment was done within a span of three weeks.

4.3 Test Methods

Prior to the training session, all the subjects were screened for color vision and visual acuity using Ishihara and Snellen charts (20/25), respectively. Please note that subjects were given instructions not to consider the ghost and stitching artifacts while rating the video quality.

4.3.1 Experiment 1

For evaluating the subjective quality, the M-ACR method was proposed by the author in Singla et al. (2017). In the M-ACR test method, test videos are shown twice because the short test sequences are only 10 s long, and subjects are not used to watching 360° content. After the second subsequent viewing of a given PVS, a five-point scale[2] (5: Excellent, 4: Good, 3: Fair, 2: Poor, 1: Bad) was displayed on the HMD Rift's screen at four azimuth angles at the same time, so as to always be visible, see Fig. 4.1. Subjects were then asked, "How would you rate the quality of this video?" For rating, the user had to say aloud the rating number so that the experimenter could note down the rating of the video. This way, the user could wear the HMD continuously throughout each of the four 15 min test sessions. Figure 4.2 shows the presentation of one stimulus.

[2] It was not intended to have a strong contrast between a PVS and the scale that is why the scale was shown on a mid-grey.

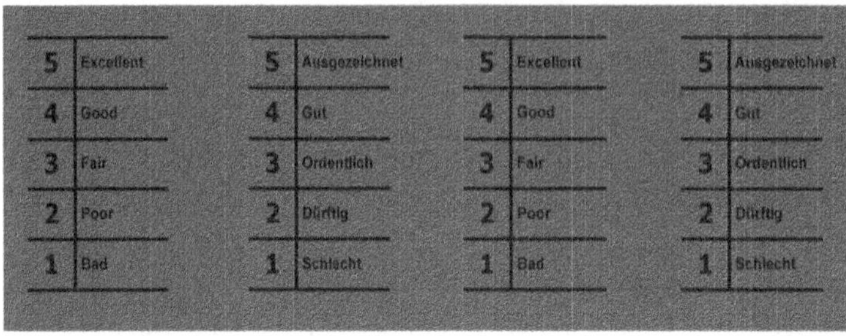

Fig. 4.1 Sample 2D representation of a 5-point Scale for ACR and M–ACR Test Methods

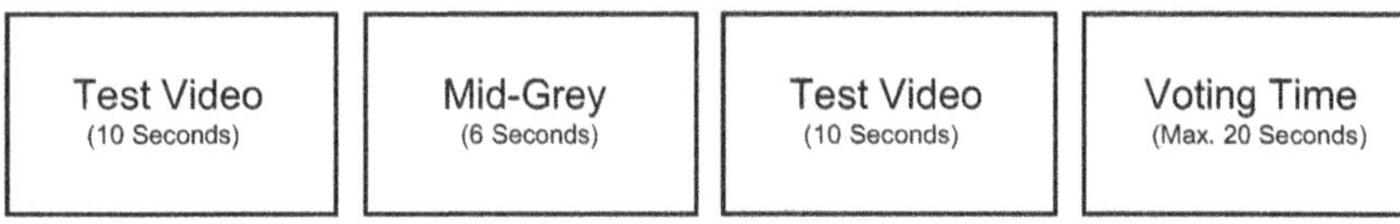

Fig. 4.2 Presentation of one stimulus for test 2 (M-ACR) (Singla et al., 2017)

The DSIS method is based on P.910 (2008) and BT.500-13 (2012). In DSIS, first, the source reference video is presented, followed by the PVS (i.e., a degraded variant of the reference video). See Fig. 4.4 for a scheme of the method. The reference videos are in raw YUV format and, hence, cannot be played using the Whirligig player. Therefore, the reference video was encoded with a high bitrate (25 Mbit/s) at 4K resolution with HEVC so that no perceivable degradations were present in the reference video.

Subjects were asked to rate the degradations on a five-point scale (5: Imperceptible, 4: Perceptible but not annoying, 3: Slightly annoying, 2: Annoying, 1: Very annoying). The scale[3] was displayed on the HMD screen at four different yaw angles, see Fig. 4.3. Subjects were asked to rate the compression artifacts and impairments. For collecting the ratings, the user had to say the number aloud and the experimenter would note down the rating. This process allowed the user to continuously wear the HMD throughout a test session.

4.3.2 Experiment 2

Three test methods, ACR (P.910, 2008), M–ACR (Singla et al., 2017), and DSIS (BT.500-13, 2012; P.910, 2008), were selected for assessing the quality of 360° videos. In ACR and M-ACR, subjects were asked to rate the absolute quality

[3] It was not intended to have a strong contrast between a PVS and the scale that is why the scale was shown on a mid-grey.

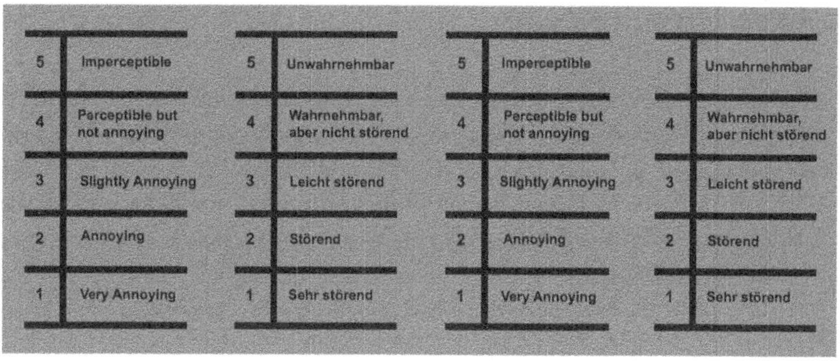

Fig. 4.3 Sample 2D representation of a 5-point Scale for DSIS Test Method

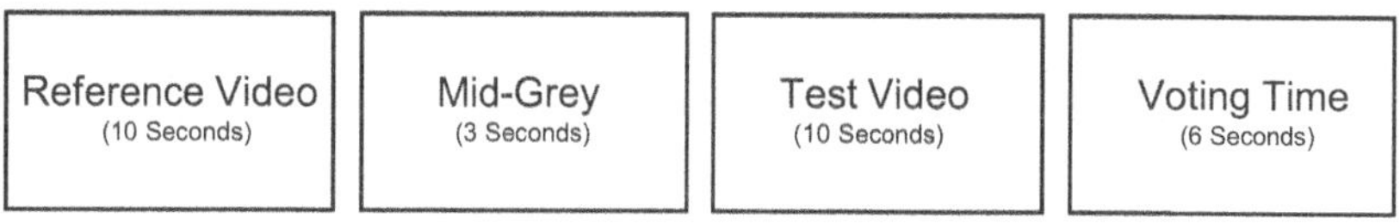

Fig. 4.4 Presentation of one stimulus for test 3 and 6 (DSIS) (Singla et al., 2018)

Fig. 4.5 Presentation of one stimulus for test 4 (ACR)

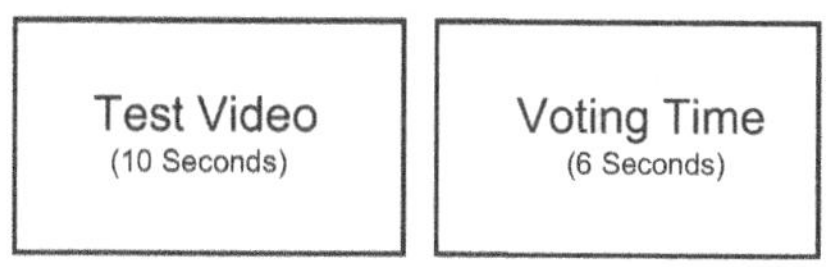

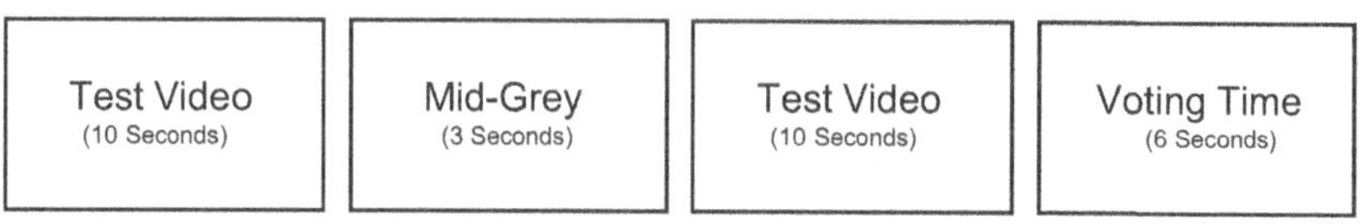

Fig. 4.6 Presentation of one stimulus for test 5 (M-ACR)

of the video on the five-point ACR scale (1: Bad, 2: Poor, 3: Fair, 4: Good, 5: Excellent). In a DSIS test, subjects were asked to rate the impairments on a five-point scale (1: Very annoying, 2: Annoying, 3: Slightly annoying, 4: Perceptible but not annoying, 5: Imperceptible). The reference video was encoded with a high bitrate (35 Mbit/s) at 8K resolution with the HEVC encoder, to ensure that there were no perceivable/noticeable degradations present. The total time for assessing one PVS for the different test methods was 16 s for ACR, 29 s for M–ACR, and 29 s for DSIS.

The procedure for the presentation of one stimulus is shown in Figs. 4.4, 4.5, and 4.6 for the DSIS, ACR, and M-ACR test methods, respectively.

4.4 Experimental Results

For analyzing the results, outlier detection was performed in order to check the reliability of the subjects. The Pearson correlation coefficient was computed between the scores of each user to the Mean Opinion Score (MOS) value. A threshold of 0.75 was considered as a reliability criterion for a given user (Tan et al., 2014). None of the subjects left the experiment in between due to simulator sickness in any of the experiments.

4.4.1 Experiment 1

For the M-ACR (Test 2) and DSIS (Test 3) test methods, only one subject was found to be an outlier in each of the test method. After outlier removal, the MOS values and associated 95% confidence intervals (CIs) were computed for each test condition (bitrates, content, and resolution) as per ITU guidelines (BT.500-13, 2012).

Figure 4.7 shows the MOS for different bitrates and for the M-ACR and DSIS methods for two different resolutions – 4K and FHD[4,5]. The motivation behind selecting 4K and FHD resolution is the resolution limitation of the Oculus Rift. Therefore, it was investigated if these two resolutions have an impact on the assessment of the subjective quality evaluation in a virtual reality system. It is evident that 4K leads to a higher perceived quality than FHD irrespective of the test method.

Figure 4.8 shows that DSIS provides a larger range of MOS values – although only slightly – and thus seems to provide a higher subjective resolution power. M-ACR and DSIS differ in the evaluation and in the presentation method: DSIS evaluates the impairments present in the video, while M-ACR evaluates the absolute quality of the video. DSIS could, therefore, be more useful for assessing fine-grained details, especially at 4K resolution. It is worth noting, however, that CI values are considerably higher in the case of the DSIS method, as seen in Fig. 4.8, where the CI corresponding to each MOS was plotted for all the PVSs.

In Fig. 4.9, the standard deviation is plotted over the respective MOS rating as described in Hoßfeld et al. (2011). To illustrate the underlying principle, the theoretical cases of a $MOS = 1$ and a $MOS = 5$ can be used. In these instances, the standard deviation must be 0, since all subjects have to agree on a rating of 1 or 5, respectively, to achieve these values as the mean. The values obtained for the medium range of the MOS-scale are the highest since here a variety of different individual ratings can lead to the respective MOS value. For the M–ACR

[4] While selecting the different bitrates for FHD resolution, it was seen that using the bitrates beyond 4.3 MBit/s will not deliver a noticeable improvement in visual quality. Hence for the experiment, the bitrate was limited to 5 MBit/s.

[5] https://netflixtechblog.com/per-title-encode-optimization-7e99442b62a2

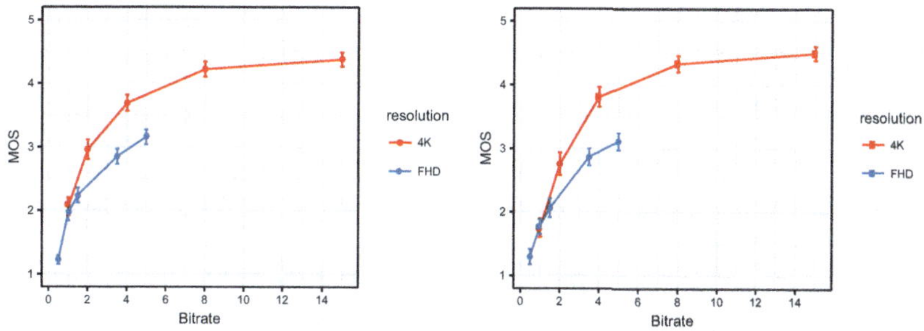

(a) (b)

Fig. 4.7 MOS with corresponding CIs for different test methods for experiment 1 (Singla et al., 2017, 2018). Please see Fig. 5.3 (Sect. 5.2), where the results are analyzed in terms of bitrate and resolution. (**a**) M-ACR (Test 2). (**b**) DSIS (Test 3)

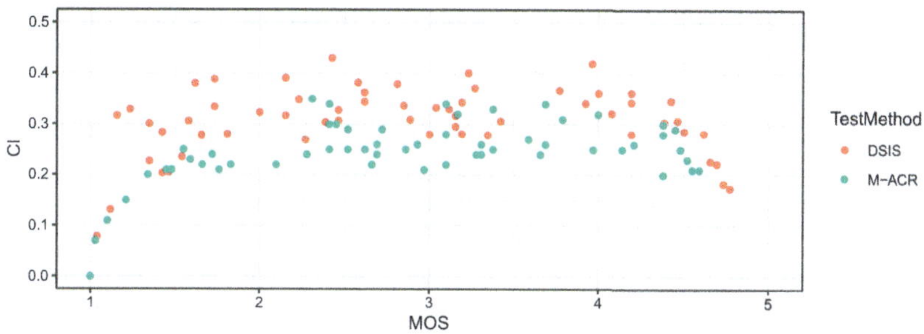

Fig. 4.8 CI vs. MOS for experiment 1 (Singla et al., 2018)

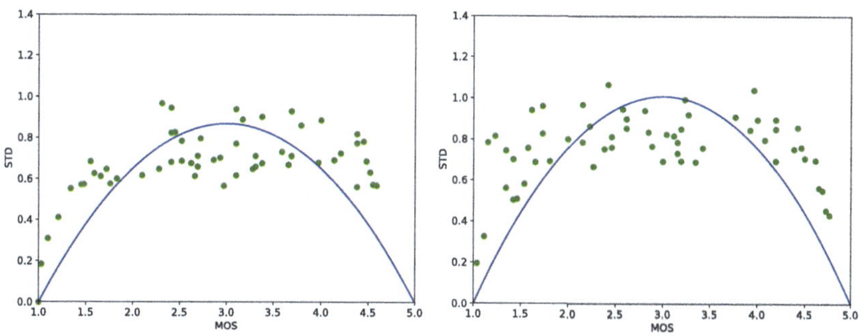

(a) (b)

Fig. 4.9 SOS hypothesis based on Hoßfeld et al. (2011) for experiment 1. (**a**) M-ACR (Test 2). (**b**) DSIS (Test 3)

and DSIS test methods, there is one MOS value that is close to one. However, the corresponding standard deviation values are not zero or close to zero. This is because if all of the subjects have given the same rating (1) to the particular PVS, except one

Table 4.1 MCI, MOS Range, and MCI_{norm} for DSIS and M-ACR for experiment 1 (Singla et al., 2018)

	DSIS	M-ACR
MCI	0.136	0.118
MOS Range	3.198	3.149
MCI_{norm}	0.042	0.037

Table 4.2 ANOVA results on MOS vs. experimental factors for M-ACR test method (Experiment 1) (Singla et al., 2017)

	DF	Sum Sq	Mean Sq	F	p
resolution	1	20.821	20.821	82.090	<0.01
bitrate	1	23.254	23.254	91.684	<0.01
content	5	7.731	1.546	6.096	<0.01
resolution × bitrate	1	4.372	4.372	17.238	<0.01
resolution × content	5	0.321	0.064	0.253	0.93
bitrate × content	5	1.590	0.318	1.254	0.30
resolution × bitrate × content	5	0.195	0.039	0.154	0.97

subject (e.g., 2). The MOS of the PVS is close to 1 (e.g., 1.04) and the standard deviation would be close to 0.2 for $N_{Subjects} = 26$.

The SOS parameter for the M–ACR and DSIS test methods is $a_{M-ACR} = 0.217$ and $a_{DSIS} = 0.252$, respectively. The a value is in the same order of magnitude, indicating that these two test methods have a strong similarity. Obviously, the Pearson correlation coefficient between these two methods is very high ($r = 0.992$).

In order to find out which evaluation methods is more reliable, the statistical reliability was evaluated based on Tominaga et al. (2010). For both the test methods, MCI_{norm} was calculated as shown in Eq. 4.1.

$$MCI_{norm} = \frac{MCI}{MOS\ Range}. \tag{4.1}$$

Here, MCI is the *Mean Confidence Interval. MOS Range* is the absolute difference between the highest and lowest MOS for each test method. Table 4.1 shows the MOS, MOS Range, and MCI_{norm} for DSIS and M-ACR methods. Results indicate that M-ACR appears slightly better than DSIS for these results, as MCI and MCI_{norm} are smaller for M-ACR. A post-hoc multiple comparisons using a Bonferroni-corrected t-test between each PVS showed that out of 1770 (60 × 59/2) possible condition pairs, 1083 condition pairs can be differentiated for the M-ACR method and 1030 in the case of DSIS, respectively. Therefore, post-hoc results support the observation that M-ACR is more reliable than DSIS for experiment 1.

In order to find the impact of the experimental factors on the users' ratings, an ANOVA (Analysis of Variance) was conducted on the MOS, with resolution, bitrate, and video sequence (content) as independent variables. As shown in Tables 4.2 and 4.3, all have a significant impact ($p < 0.01$), and there is an interaction effect between resolution and bitrate for the M-ACR and DSIS test methods.

Table 4.3 ANOVA results on MOS vs. experimental factors for DSIS test method (Experiment 1) (Singla et al., 2018)

	DF	Sum Sq	Mean Sq	F	p
resolution	1	21.878	21.878	64.244	<0.001
bitrate	1	30.900	30.900	90.739	<0.001
content	5	7.980	1.596	4.687	<0.01
resolution × bitrate	1	3.794	3.794	11.142	<0.01
resolution × content	5	0.523	0.105	0.307	0.90
bitrate × content	5	1.318	0.264	0.774	0.57
resolution × bitrate × content	5	0.147	0.029	0.086	0.99

4.4.2 Experiment 2

For the tests with the DSIS, ACR, and M–ACR test methods 3, 4, and 5 subjects were found to be outliers, respectively. As a result, at least 24 individual scores for each test method are being considered. Figure 4.10 shows the MOS averaged over all video sequences with the corresponding 95% CI for all three test methods. One of the main research goals was to investigate whether subjects are able to detect differences in perceived quality provided by 4K, 6K, and 8K resolutions. Results indicate that subjects are able to detect differences in perceived quality provided by 4K and 6K resolutions irrespective of the rating method. The advantage of using 8K over 6K cannot clearly be observed from the results for all the test methods. The possible reason for this behavior may result from the resolution limitation of the HTC Vive Pro HMD (2880 × 1600) used in this experiment.

From Fig. 4.10 it is obvious that DSIS provides a higher dynamic range in ratings when compared to the other two evaluation methods (ACR and M–ACR), covering almost the entire scale range. Therefore, based on experiment 2, DSIS could be useful to evaluate fine-grained details in the video, especially at higher resolutions.

Another observation relates to the general shape of the quality–bitrate-dependency curves, see Fig. 4.10. For DSIS, the result implies that both bitrate and resolution contribute to the ratings. For ACR, this is also the case, however, the resolution seems to play less of a role. For M-ACR, in turn, the resolution appears to not play any role: all result curves lie on top of each other. A possible explanation is that in M-ACR, with its double presentation, there is an adaptation to the current condition, reducing the impact of finer differences between conditions. In turn, for ACR and DSIS, subjects can better rely on some inter-stimulus comparisons. For DSIS, there is a direct comparison with a reference, and for ACR there are the typical sequence effects, letting subjects compare with the just preceding stimulus. This effect will depend on the stimulus duration, and with the short 10 s-stimuli used in the tests, such sequential effects may still be present.

Figure 4.11 shows the STD plotted over MOS for all rating methods as described in Hoßfeld et al. (2011). The SOS parameter for the ACR, M–ACR, and DSIS test methods is $a_{ACR} = 0.219$, $a_{M-ACR} = 0.240$ and $a_{DSIS} = 0.237$, respectively. The

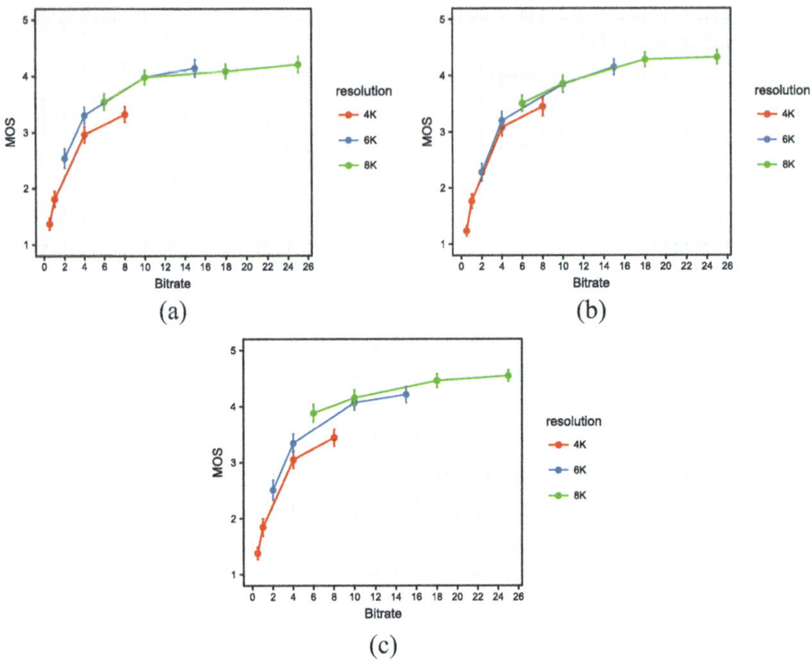

Fig. 4.10 MOS with corresponding CIs for different test methods for experiment 2 (Singla et al., 2019). Please see Fig. 5.4 (Sect. 5.2), where the results are analyzed in terms of bitrate and resolution. (**a**) ACR (Test 4). (**b**) M–ACR (Test 5). (**c**) DSIS (Test 6)

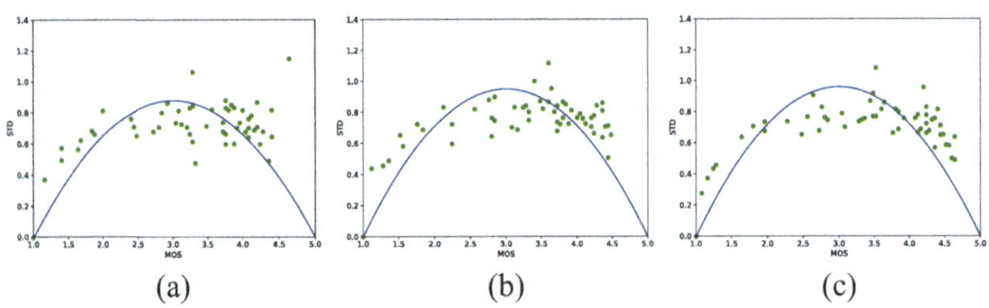

Fig. 4.11 SOS hypothesis based on Hoßfeld et al. (2011) for experiment 2. (**a**) ACR (Test 4). (**b**) M-ACR (Test 5). (**c**) DSIS (Test 6)

a value is in the same order of magnitude, indicating that these three test methods have a strong similarity.

In Fig. 4.11 also there are a couple of MOS values that are close to one for the M–ACR and DSIS test methods. Here too, the corresponding standard deviation values are not zero or close to zero. This is because if all of the subjects have given the same rating (1) to the particular PVS, except one subject (e.g., 2). The MOS of the PVS is close to 1 (e.g., 1.04) and the standard deviation would be close to 0.2.

Table 4.4 MCI, MOS Range, and MCI_{norm} for ACR, M-ACR, and DSIS for experiment 2 (Singla et al., 2019)

Test method	ACR	M-ACR	DSIS
MCI	0.138	0.1392	0.1405
MOS Range	2.824	3.088	3.16
MCI_{norm}	0.0489	0.0450	0.0444

Table 4.5 Pearson correlation coefficient between different test methods (Singla et al., 2019)

Test method	ACR	M-ACR	DSIS
ACR	X	0.95	0.93
M–ACR	0.95	X	0.97
DSIS	0.93	0.97	X

To investigate which rating method is more reliable, statistical reliability was computed, as described in Tominaga et al. (2010). To this aim, the MCI_{norm} was calculated according to Eq. 4.1. Table 4.4 shows the MCI, MOS range, and MCI_{norm} for all test methods. Results suggest that DSIS is statistically more reliable than the other two test methods (ACR and M–ACR). A post-hoc multiple comparisons using a Bonferroni-corrected t-test between each PVS showed that out of 1770 ($60 \times 59/2$) possible condition pairs, 885 condition pairs can be differentiated for the DSIS method and 808 and 806 in the case of ACR and M–ACR, respectively. Therefore, post-hoc results for experiment 2 support the observation that DSIS is more reliable than ACR and M–ACR.

The Pearson Correlation Coefficient between the mean values obtained with the different test methods was further analyzed. Table 4.5 shows the correlation coefficient between different test methods. Results suggest that all three test methods are highly correlated, especially M–ACR and DSIS (0.97).

To statistically evaluate the impact of experimental factors (resolutions, bitrates, and video sequences) on the ratings, an ANOVA was carried out on the MOS, with resolution, bitrate, and video sequence (content) as independent variables. Results show that resolutions and bitrates all have a statistically significant contribution, with p-values <0.01, and there is an interaction effect, which is significant, between resolution and bitrate irrespective of the test method as shown in Tables 4.6, 4.7, and 4.8. For the DSIS and ACR test methods, content has a statistically significant contribution, with p-values <0.01, and there is an interaction effect (statistically significant) between bitrate and content.

4.5 Discussion

The results of video quality obtained from experiments 1 and 2 were compared. In experiment 1, the video quality of 360° videos was compared using the M–ACR and DSIS test methods for 4K and FHD resolutions. Based on the evaluation results, M–ACR has been found to be providing results with resolving power superior over DSIS, based on statistical reliability and post-hoc analysis. However, the results

Table 4.6 ANOVA results on MOS vs. experimental factors for ACR test method (Experiment 2)

	DF	Sum Sq	Mean Sq	F	p
resolution	2	26.398	13.199	109.819	<0.01
bitrate	1	10.503	10.503	87.387	<0.01
content	4	2.582	0.645	5.370	<0.01
resolution × bitrate	2	8.483	4.241	35.289	<0.01
resolution × content	8	0.648	0.081	0.674	0.71
bitrate × content	4	2.607	0.652	5.422	<0.01
resolution × bitrate × content	8	0.754	0.094	0.784	0.620

Table 4.7 ANOVA results on MOS vs. experimental factors for M-ACR test method (Experiment 2)

	DF	Sum Sq	Mean Sq	F	p
resolution	2	26.339	13.169	90.047	<0.01
bitrate	1	15.343	15.343	104.909	<0.01
content	4	1.918	0.480	3.279	0.02
resolution × bitrate	2	9.961	4.981	34.056	<0.01
resolution × content	8	0.305	0.038	0.261	0.97
bitrate × content	4	1.738	0.435	2.971	0.0352
resolution × bitrate × content	8	0.601	0.075	0.514	0.8363

Table 4.8 ANOVA results on MOS vs. experimental factors for DSIS test method (Experiment 2)

	DF	Sum Sq	Mean Sq	F	p
resolution	2	33.87	16.934	132.751	<0.01
bitrate	1	12.31	12.315	96.540	<0.01
content	4	4.49	1.123	8.805	<0.01
resolution × bitrate	2	9.21	4.606	36.107	<0.01
resolution × content	8	0.75	0.094	0.738	0.65
bitrate × content	4	2.35	0.586	4.597	<0.01
resolution × bitrate × content	8	0.65	0.081	0.637	0.740

may have been impacted by the seasonal weather conditions when conducting the respective studies: the M–ACR test was conducted in winter and the DSIS test in summer, during a period of particularly hot weather, possibly affecting participants, especially during their arrival to the test lab. The outside conditions may have therefore biased the quality ratings and the SSQ scores, leading to reduced resolving power. In an informal internal test, it was also found that outside temperature has an effect on the scores of simulator sickness, where, e.g., symptoms such as sweating, fatigue, etc., are rated higher. Therefore, to decrease the contextual effect on the experimental results, the study presented in experiment 2 was conducted under the same contextual conditions. Based on the results presented for experiment 2, DSIS was found to be statistically more reliable and with more resolving power than M–ACR and ACR. It is considered that the results obtained in experiment 2 are more reliable than those presented in experiment 1, since in experiment 2 the comparisons between ACR, M–ACR, and DSIS were conducted under the same contextual conditions.

4.6 Conclusions

In this chapter, the perceived quality of 360° videos was evaluated and compared using different test methods in several experiments for different resolutions and bitrates. Different statistical properties of the results to assess the statistical reliability and resolving power of the different methods were analyzed. Results based on experiment 2 indicate that DSIS is statically more reliable and provides more resolving power than ACR and M–ACR. Based on the assessment time and statistical reliability, it is recommended to use the DSIS test method for evaluating the subjective quality of 360° video playback in high resolutions, when a more fine-grained resolving power is sought. In case an absolute rating is sought, a single-stimulus method such as the ACR can be used, when somewhat less resolving power of the method is acceptable and the testing time is of relevance.

In future work, more subjective tests need to be conducted to include other test methods, for example, paired comparison and SAMVIQ.[6]

References

ITU-R BT.500-13, *Recommendation ITU-R BT.500-13: methodology for the subjective assessment of the quality of television pictures* (2012)

S. Fremerey et al., AVtrack360: an open dataset and software recording people's head rotations watching 360-degree; videos on an HMD, in *Proceedings of the 9th ACM MMSys*, Amsterdam (2018). ISBN: 978-1-4503-5192-8. https://doi.org/10.1145/3204949.3208134

T. Hoßfeld, R. Schatz, S. Egger, SOS: the MOS is not enough!, in *2011 Third International Workshop on Quality of Multimedia Experience* (2011), pp. 131–136. https://doi.org/10.1109/QoMEX.2011.6065690

R.S. Kennedy et al., Simulator sickness questionnaire: an enhanced method for quantifying simulator sickness. Int. J. Aviat. Psychol. **3**(3), 203–220 (1993)

ITU-T Rec. P.910, *Recommendation ITU-T P.910: Subjective video quality assessment methods for multimedia applications* (2008)

ITU-T P.919, *Subjective test methodologies for 360° video on head-mounted displays*. Recommendation P.919 (2020)

[6] In this test method, test participants are allowed to watch a PVS multiple times and can change their votes at different points in time. A continuous scale is used from 0 to 100 with five labels (excellent, good, fair, poor, bad).

A. Singla et al., Comparison of subjective quality evaluation for HEVC encoded omnidirectional videos at different bit-rates for UHD and FHD resolution, in *Proceedings of the on Thematic Workshops of ACM Multimedia*, Mountain View (2017). ISBN: 978-1-4503-5416-5. https://doi.org/10.1145/3126686.3126768

A. Singla, W. Robitza, A. Raake, Comparison of subjective quality evaluation methods for omnidirectional videos with DSIS and Modified ACR. Electron. Imaging **2018**, 1–6 (2018). https://doi.org/10.2352/ISSN.2470-1173.2018.14.HVEI-525

A. Singla, W. Robitza, A. Raake, Comparison of subjective quality test methods for omnidirectional video quality evaluation, in *2019 IEEE 21st International Workshop on Multimedia Signal Processing (MMSP)* (2019), pp. 1–6

T.K. Tan et al., HEVC subjective video quality test results, in *IBC2014 Conference, 2014*, Amsterdam, 11–15 Sept 2014 (Institution of Engineering and Technology, UK, 2014). ISBN: 978-1-84919-927-8. https://doi.org/10.1049/ib.2014.0029

T. Tominaga et al., Performance comparisons of subjective quality assessment methods for mobile video, in *2010 Second International Workshop on QoMEX* (2010), pp. 82–87. https://doi.org/10.1109/QOMEX.2010.5517948

Chapter 5
Visual Quality Assessment for 360° Videos

5.1 Impact of HMDs

The test 1 dataset is used to compare the integral quality of the two HMDs, namely, HTC Vive and Oculus Rift, and is described in Sect. 3.2.1.

5.1.1 Technical Setup and Equipment

The subjects were asked to watch the omnidirectional videos with two HMDs – HTC Vive and Oculus Rift (Consumer Version 1) – in order to evaluate the integral quality of the VR presentation. In order to decrease the contextual effect, the names of both HMDs were hidden and replaced by arbitrary identifiers. The resolution and field of view of both devices are 2160 × 1200 and 110°, respectively.

Only one subject performed the test at a given time. For hygienic reasons, after every test subject session, the HMDs and their lenses were cleaned. Each subject had to rate 24 360° sequences (6 contents, 2 resolutions, and 2 HMDs, full-factorial within-subject design). The subjects had to first watch the video for 60 seconds, take off the device, and then had about 60 seconds to rate the video and fill in the simulator sickness questionnaire.[1] Then they were given 60 seconds of pause in order to minimize the effect of one video over another on quality ratings, simulator sickness, and behavior.

The selection of video sequences (Test 1), description of the test environment, and the test software have been explained in Chap. 3.

[1] Before the start of the test, subjects were asked to read the SSQ multiple times to familiarize themselves with it.

A. Singla, *Assessment of Visual Quality and Simulator Sickness for Omnidirectional Videos*, T-Labs Series in Telecommunication Services,
https://doi.org/10.1007/978-3-031-49988-3_5

5.1.2 Test Method

For the assessment of the integral quality of 360° videos, the ACR test method was selected (P.910 2008). In an ACR test, videos are shown only once and subjects are asked to rate the quality on a five-level scale (5: Excellent, 4: Good, 3: Fair, 2: Poor, 1: Bad). The subjects were given instructions not to consider stitching and ghost artifacts while rating the video. Dedicated individual playlists were used for each subject, so as to systematically avoid presentation order effects.

In total, 28 subjects participated in the subjective test. Out of 28, 15 were females and 13 were males. The range of the subjects' age varied from 20 to 38, with an average of 26.25 and a median of 25 years.

All the subjects were screened for correct visual acuity using Snellen charts (20/20) and for color vision using Ishihara charts. Prior to the experiment, written instructions were given to the subjects explaining their tasks. Five training samples were used in two different HMDs to familiarize the subjects with the test procedure and to adjust the HMDs according to their head size.

5.1.3 HTC Vive vs. Oculus Rift

In order to check the reliability of the subjects, outlier detection was performed on the raw scores according to ITU guidelines in BT.500-13 (2012). In this test, none of the subject was detected as an outlier, which led to 28 individual scores for each stimulus being considered in the results.

Choosing the 4K and FHD resolution was motivated by the resolution limitations of the HMDs' displays. Hence, one goal was to investigate whether the users would be able to notice the difference between these two resolutions in both devices or not, and which of the devices provide better subjective video quality. Figure 5.1 shows the MOS of HTC Vive and Oculus Rift for 4K and FHD. It is evident from

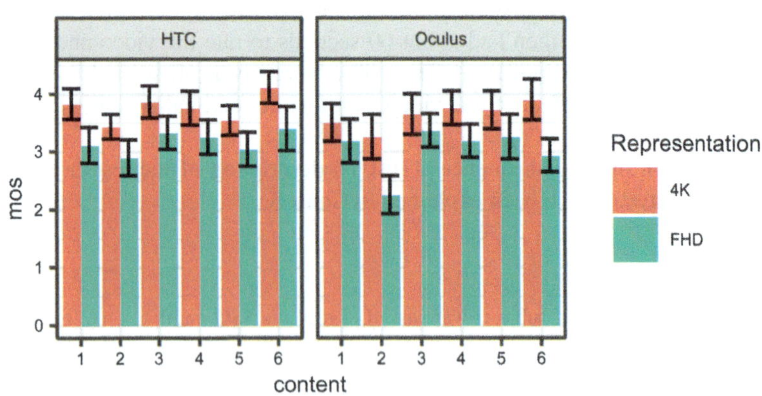

Fig. 5.1 MOS with CIs for all devices and representations for test 1 (Singla et al. 2017c)

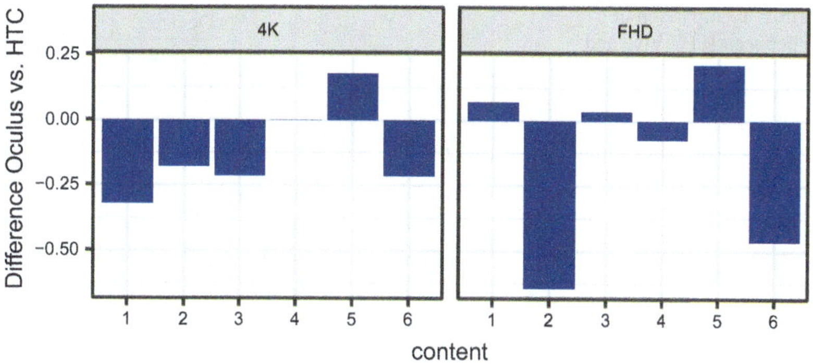

Fig. 5.2 Difference in the MOS between devices for different representations for test 1 (Singla et al. 2017c)

the plots that the subjects found differences between the resolution of the contents irrespective of the device and content. It is noted that not all differences between contents were statistically significant, as can also be seen from the plot in Fig. 5.1. Obviously, some source videos were better suited to convey resolution and bitrate differences than others. Figure 5.2 shows the difference between the MOS values of HTC Vive and Oculus Rift for 4K as well as FHD. Positive bars mean that Oculus Rift is better than HTC Vive. For the 4K resolution, it can clearly be seen that HTC Vive is preferred over Oculus Rift, except for content 5. For FHD resolution, Oculus is preferred for contents 1, 3, and 5, but the difference between the MOS of Oculus Rift and HTC Vive is less. A paired t-test was performed on the subjects' ratings for these three sequences, showing that the difference is not significant ($p = 0.408$). For contents 2 and 6, HTC Vive is preferred, and the difference between MOS values is also considerable. Here too, a paired t-test was performed on the subjects' ratings for these two sequences, showing that the difference is extremely significant ($p = 0.0003$). Table 5.1 shows the MOS averaged over all the contents for both resolutions and HMDs. It is clear from the values that HTC Vive leads to a slightly higher MOS for 4K as well as FHD resolution than Oculus Rift. The results imply that HTC Vive provides slightly better integral quality than Oculus Rift, although both the HMDs have the same technical specifications.[2] The difference in perceived quality could be due to the rendering algorithm or type of lenses used. A paired t-test was performed on the subjects' ratings for these two devices, showing that the difference is not statistically significant for 4K ($p = 0.0980$) resolution and FHD ($p = 0.1090$) resolution at a confidence level of 95%.

In order to statistically evaluate the impact of the experimental factors (HMDs, resolutions, and contents) on the subjects' ratings, a within-subjects ANOVA was carried out on the individual ratings. The results (see Table 5.2) suggest that

[2] The FOV, refresh rate, and resolution are the same for Rift and Vive.

Table 5.1 Comparison of MOS between HTC Vive and Oculus Rift for test 1 (Singla et al. 2017c)

Device	4K	FHD
HTC Vive	3.75	3.16
Oculus Rift	3.62	3.02

Table 5.2 Results from within-subjects ANOVA on ratings for test 1 (Singla et al. 2017c)

	DF	Sum Sq	Mean Sq	F	p
Device	1	3.01	3.013	3.48	0.07
Resolution	1	58.93	58.93	58.33	<0.01
Content	5	28.86	5.77	7.36	<0.01
Device × resolution	1	0.01	0.01	0.04	0.85
Device × content	5	6.71	1.34	2.13	0.07
Resolution × content	5	4.08	0.82	2.08	0.07
Device × resolution:content	5	3.50	0.70	1.56	0.17

resolution and content have a statistically significant contribution to the ratings, with all p-values <0.01. The device itself has a slight effect ($p = 0.07$) that is not significant at 5% level. In addition, there is a slight interaction between content and device ($p = 0.07$) as well as resolution and content ($p = 0.07$), also not significant at 5% level.

5.2 Encoding Settings

In Test 1, as presented in Sect. 5.1, the resolution is jointly factored with bitrate. To more systematically investigate the impact of resolution and bitrate, the results of additional Tests 2–6 are analyzed in the following section. To this aim, this section presents the results for the assessment of the subjective quality of HEVC-encoded videos for different resolutions, 8K (7680 × 3840), 6K (6144 × 3072), 4K (3840 × 1920) and FHD (1920 × 1080) at different bitrates, bearing in mind the limitation of the resolution of the employed HMDs (Oculus Rift (2160 × 1080) and HTC Vive Pro (2880 × 1600)). *libx265* (HEVC) was used to encode the test sequences in order to achieve a set of target bitrates. The information about the bitrates is given in Tables 3.4 and 3.5.

A series of 360° video quality tests (Tests 2–6) was conducted using different test methods (ACR, M–ACR, and DSIS). The test conditions were chosen so as to reflect a typical use case such as finding the optimal balance between the available network resources and perceived quality. Thus, the test results provide insights into video quality for different resolution and encoding settings ("bitrate ladders"). These may be useful for choosing appropriate representations in the context of HTTP-based adaptive streaming in the case of full frame streaming instead of tile-based streaming, as it is discussed in Sect. 5.3.

Table 5.3 Test conditions for test 8 (Singla et al. 2019b)

Resolution	Storage	Delay and Bandwidth settings				
		0 ms	12 ms	47 ms	112 ms	Maximum Bandwidth (in Mbit/s)
4K	Server		X	X	X	8
6K	Server		X	X		20
8K	Server		X	X	X	25
4K (Anchor)	Phone	X				8
6K (Anchor)	Phone	X				20

Table 5.4 Test conditions for test 9 (Singla et al. 2019b)

Resolution	Storage	Delay and Bandwidth settings				
		0 ms	12 ms	47 ms	112 ms	Maximum Bandwidth (in Mbit/s)
TBD[a]	Server		X	X		10
4K, 8K, TBD[a]	Server		X	X	X	15
TBD[a]	Server		X	X		25
6K (Anchor)	Phone	X				40

[a] Resolution decided by streaming solution provider

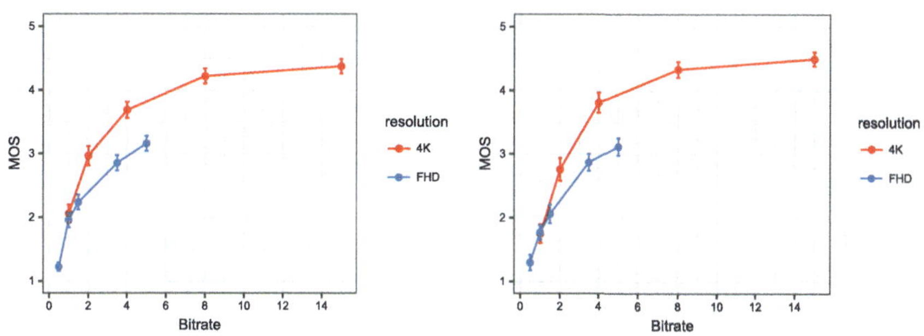

(a) (b)

Fig. 5.3 MOS with corresponding CIs for different test methods for test 2 and 3 (Singla et al. 2017b, 2018). Please see Fig. 4.7 (Sect. 4.4.1), where the results are analyzed on the reliability and validity of the test methods. (**a**) M-ACR (Test 2). (**b**) DSIS (Test 3)

The selection of video sequences, description of the test environment, and the test software were explained in Chap. 3. The test methods used in the tests were explained in Chap. 4.

Figure 5.3 shows the MOS averaged over all the contents for different bitrates. It is evident that 4K leads to a higher perceived quality than FHD irrespective of the test method. Additionally, the perceived quality at 15 MBit/s is only slightly higher than at 8 MBit/s for 4K resolution. To reduce the bandwidth requirements and to ensure an optimal network delivery, for this particular instance of HEVC encoding and full frame delivery, it is advisable to send the content at 8 MBit/s rather than at

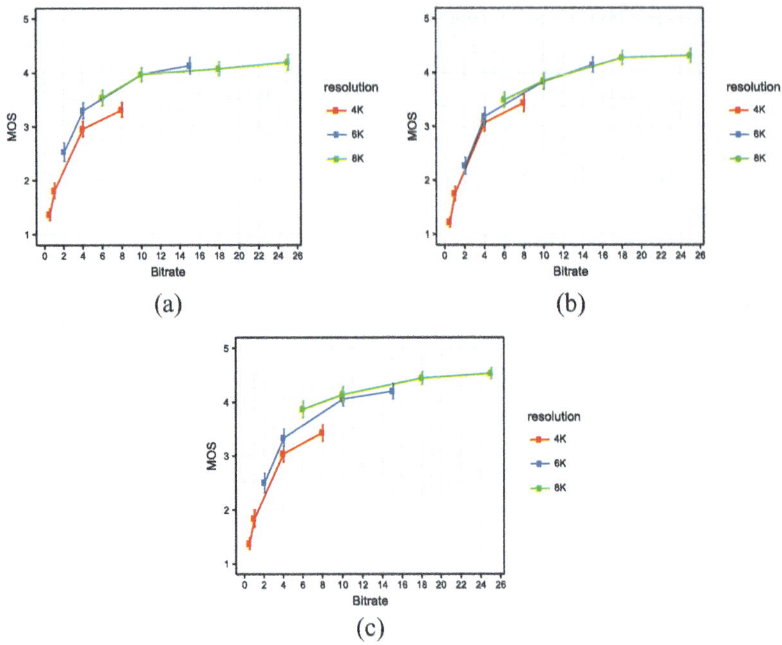

Fig. 5.4 MOS with corresponding CIs for different test methods for test 4, 5, and 6 (Singla et al. 2019a). Please see Fig. 4.10 (Sect. 4.4.2), where the results are analyzed on the reliability and validity of the test methods. (**a**) ACR (Test 4). (**b**) M-ACR (Test 5). (**c**) DSIS (Test 6)

15 MBit/s at a marginal cost in the perceived quality. This will help in saving almost 50% bandwidth, thus enabling efficient distribution of 360° contents.

Figure 5.4 shows the MOS averaged over all video sequences with the corresponding 95% CI for all three test methods. Results indicate that 6K provides better-perceived quality over 4K irrespective of the rating method. Additionally, the advantage of using 8K resolution over 6K cannot be clearly seen for any of the three test methods. This could be possible due to the resolution limitation of the HTC Vive Pro HMD. In terms of the technology settings underlying identifiable differences, it can be stated that the perceived quality at 15 Mbit/s for 6K is marginally lower than 25 Mbit/s for 8K for all test methods. Therefore, if 360° videos are encoded at around 15 Mbit/s at 6K resolution for full frame (instead of tile-based) streaming, there is only a marginal loss in perceived quality in relation to the high-quality reference condition used. This will help in saving almost 40% bandwidth, thus enabling efficient distribution of 360° contents.

Please note that the findings that are reported in this Sect. 5.2 are valid only for the specifics type of HMDs such as Oculus Rift, HTC Vive, and HTC Vive Pro or for another HMD that has a similar resolution to these aforementioned HMDs. If an HMD (HTC Vive Pro 2) has a higher resolution (4896×2448), the above result may not hold true.

5.3 Tile-Based vs. Full Frame Streaming

In this section, a testbed and a test method for QoE tests of viewport-adaptive streaming approaches are proposed. The proposed testbed allows to evaluate 360° video streaming solutions using different test conditions. The subjective tests were conducted to validate the testbed and investigated the influence of resolution, bandwidth, and network round-trip delay on QoE aspects of tile-based streaming of 360° videos.

The proposed method is validated for two different scenarios: (1) Resolution- and Bandwidth-limited, (2) Bandwidth-limited only, for different simulated network latency values, using a specific tile-based streaming technology available in the market. The chosen input variables (video sequences, resolution, bandwidth, and network round-trip delay) are tested for their statistical significance.

5.3.1 Experimental Setup

In this experiment, two tests were conducted; both used the same video sequences. Two different test setups were used:

- Test 8: Resolution- and Bandwidth-limited
    1. There were two anchor conditions: one having 4K and the other with 6K resolution, with bandwidth limits of 8 and 20 Mbit/s, respectively.
    2. The number of *Hypothetical Reference Circuits* (HRCs) per video was 10.
    3. Each test session was 10 minutes long.
- Test 9: Bandwidth-limited
    1. Only one anchor condition was used: having 6K resolution with a maximum of 40 Mbit/s.
    2. The number of HRCs per video was 14.
    3. The duration of each test session was 14 minutes.

It was decided to leave out the audio of the tests, as the goal was to evaluate subjective video quality. Since audio quality can affect video quality perception, this would have been an undesirable influence on the results (Beerends and De Caluwe 1999). Note that the influence of video quality on perceived audio quality is higher than the other way around, but audio quality *does* affect video quality and this influence should be avoided.

The proposed test method is validated in both the above test setups, using ClearVR[3] tile-based streaming technology available in the market. The impact of

[3] ClearVR is an android application developed by Tiledmedia for the tile-based streaming of 360° video.

chosen input variables (video sequences, resolution, bandwidth, and network round-trip delay) are tested for their statistical significance.

The provider (Tiledmedia) of the tile streaming technology (ClearVR) prepared the test sequences using their own tooling chain, where the author sets the bandwidth limits. The provider encoded the sequences such that they stayed within those previously announced bandwidth limits. This was validated using network monitoring tools as described below. The processed sequences were then transferred to the local server. The provider also created a special version of its player software, with all branding removed. All network traffic was monitored to verify that the content was actually streamed during the tests and not cached locally on the devices.

The objective was to derive a method of testing systems that can be treated as black boxes. The author gave constraints (the bitrate conditions), and the system under test was to perform its best within those constraints. This is a fair way to evaluate competing systems. The section details how such an evaluation can be conducted, and then also provides the results of the evaluation.

5.3.2 *Testbed*

It was decided to have an independent, local server setup during the tests to ensure the least possible external influencing factors. The testbed had no access to the Internet and could not send requests to other services. Figure 5.5 presents the technical setup. In general, four parts are involved in the setup:

1. Web and DNS server
2. Network shaping and monitoring computer (with two network cards)
3. Wireless access point
4. Head-mounted display based on a smartphone

All servers were running Ubuntu 18.04 and had an SSD and a modern CPU (Intel i7, 4550). The web server and network shaping server were cable-connected. Also, the network shaping server was connected to the wireless access point with

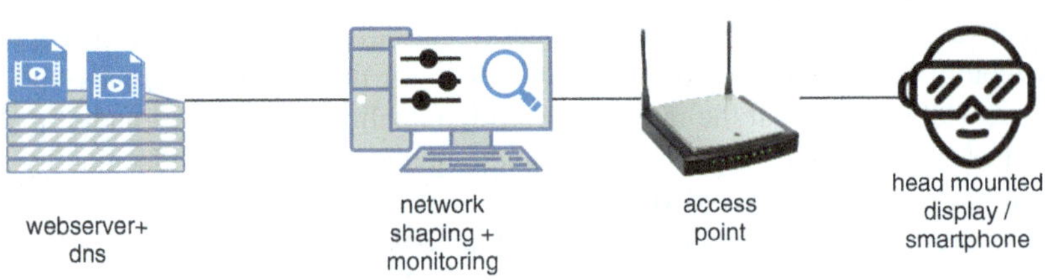

Fig. 5.5 Overview of the technical setup. The web server provides the videos, network shaping to simulate delay. The access point and smartphone deliver the content to the user (Singla et al. 2019b)

a cable. For the wireless access point, D-Link DAP-2695 was used. The HMD was connected via the 5 GHz band. A channel was selected with the least overlap with other visible access points in the test area to reduce any interference. For securing the setup, WPA2-PSK was used. The test room was in a university building with very few other WiFi access points visible, and it is ensured that there was no WiFi interference that could have affected the transmission. There are 6 WiFi SSID operating on 2.4 GHz and 2 WiFi SSID operating on 5 GHz at the location of our experiment. There should not be any interference between channel 140 and channel 64 in the 5 GHz WiFi spectrum, as they are not adjacent channels. In the 5 GHz band, each channel is 20 MHz wide, and channels are spaced 4 or 8 channels apart depending on the regulatory domain. In the case of channel 140 and channel 64, they are 16 channels apart, so there should not be any overlap or interference between them.

Apache 2.4.29[4] was used as the web server and Dnsmasq 2.79[5] for DNS, both being self-configured to replicate a realistic setup. The web server served files via HTTPS, for which a valid certificate from Let's Encrypt[6] was obtained. The network shaping used tcconfig 0.21.0,[7] which is a python-based wrapper for TC. For network monitoring, Wireshark 2.6.3[8] was used to sniff the traffic during the conducted tests. To evaluate the network performance and validity of the system, iPerf (Tirumala et al. 2005) was used to measure the maximum network throughput. The accuracy of the simulated network (RTT[9]) latency could be affected by the intrinsic delay in the server-device setup. Therefore, ConnectBot v1.9.2[10] was used to measure the "0" simulated network (RTT) latency.

For viewing the 360° videos using an HMD, Samsung Gear VR with a Samsung Galaxy S8 phone running Android was used. Before each test session, all background applications were stopped on the smartphone, which would otherwise try to communicate to external (e.g., Google) services and might decrease the bandwidth available to the test streams.

The resolution of the mobile device was 2960 × 1440 and the FOV of the HMD was 101° horizontally and 90° in the vertical direction. The Moon VR player was used to play the anchor videos in full resolution. For the play-out of the omnidirectional streaming, the ClearVR App was used. The test subjects were asked to sit on a revolving chair so that they could rotate and watch the video in any direction. After each test session, the HMD and lenses were cleaned for hygiene purposes. Two HMD devices with the exact same specifications were used, and precautions were taken to prevent the subjects from being able to discern which

[4] https://httpd.apache.org/ (Accessed on: 2018/06/03).

[5] http://www.thekelleys.org.uk/dnsmasq/doc.html (Accessed on: 2018/06/03).

[6] https://letsencrypt.org/ (Accessed on: 2018/06/03).

[7] https://github.com/thombashi/tcconfig (Accessed on: 2018/06/03).

[8] https://www.wireshark.org/ (Accessed on: 2018/06/18).

[9] RTT is the time it takes to transmit a package from sender to receiver and back.

[10] https://connectbot.org/ (Accessed on: 2018/06/18).

device and which of the two playback applications they were using at any given point in time.

5.3.3 Test Setup

The combinations of delay settings, resolutions, and content storage used in Test 8 are shown in Table 5.3.

The resolutions (4K) were chosen because they reflect what was commonly used at the time of tests, and what was possible with viewport-adaptive streaming (6K and 8K). Here, one aim was to study if users can tell the difference between 6K and 8K resolution.

The RTT network delays were chosen at 0, 35, and 100 ms. The first condition reflects a very good network with an edge server close to the user. The second condition (35 ms) reflects a realistic RTT delay in common user settings. The 100 ms condition was deliberately chosen as very high; this condition is almost never seen in practice, but it allowed studying the effect of such network delays. Table 5.3 shows delays that are 12 ms longer than these conditions, because the intrinsic delay in the server-device setup was 12 ms, to which the simulated RTT delay was added. The local condition, with content playing directly from the device, had no delay at all.

These experiments studied the effect of motion-to-high-resolution latency of a given viewport-adaptive streaming solution (ClearVR) on perceived quality. Motion-to-photon latency was determined by the devices used; the technology under evaluation did not add to this latency. In order to simulate a real-world network, an additional delay was added and thus increased the RTT to affect the motion-to-high-resolution latency.

The bitrates were chosen to reflect common access conditions in the developed world, and to determine if tiled streaming indeed lives up to its promise to reduce bitrates. For comparison, a realistic bitrate to deliver the best 4K quality for non-tiled 360° video was found to be at least 8 Mbit/s, see Sect. 5.2.

Each subject had to rate 30 PVS (3 video sequences and 10 test conditions). For Test 9, only the bandwidth was fixed and the choice of resolution was left to the streaming solution provider. The different delay settings corresponding to different bandwidth settings are shown in Table 5.4. In this test, each subject had to rate 42 PVSs (3 video sequences and 14 test conditions).

At any given time, only one subject was performing the test. First, the subject had to watch the 360° video in an HMD for 30 s, take off the device, and then had 10 s to rate the video quality and fill in the short simulator sickness questionnaire.[11] Then they were given a pause for 20 s to reduce the effect of one video over the next.

Long exposure to the VE may have a negative impact on a user's comfort, and some of the users may experience symptoms of simulator sickness. Therefore, the

[11] The results for simulator sickness are given in Chap. 6.

test was divided into three test sessions. The duration of each test session was 10 minutes for Test 8 and 14 minutes for Test 9. Subjects were asked to fill in the SSQ (Kennedy et al. 1993) before and after each test session.

Because the study sought to limit the total test time for participants, it was unfortunately not possible to have a full design in which all condition permutations were tested with all participants. The conditions in Tables 5.3 and 5.4 were chosen such that it was still possible to draw meaningful, statistically relevant, conclusions.

5.3.4 Test Method

The duration of the test videos was 30 s so that users had sufficient time to adjust to the 360° viewing situation. In the employed ACR test method, test videos were shown once and subjects were asked to rate the video quality on an 11-point scale (P.910 2008). An 11-point scale was chosen over the 5-point scale for providing higher discriminative power. Test subjects were given instructions not to consider possible stitching and ghost artefacts while rating the video quality. All subjects were screened for color vision using Ishihara charts (Ishihara 1997) and for visual acuity using Snellen charts (20/25) (P.915 2016). After screening, written instructions were given explaining the test-related tasks to the subjects. A training session with 4 different training samples was used for familiarizing the subjects with the assessment procedure and for adjusting the HMD to their head size. To avoid presentation order effects, dedicated, randomized playlists were used for each subject.

In Test 8, 28 subjects (14 males and 14 females) participated with an average age of 25.07 years and a median of 25 years. In Test 9, 25 subjects (13 males and 12 females) participated with an average age of 25.36 years and a median of 25 years. 9 Subjects participated in both of the tests.

For recording the users' responses, a paper-based scale was used after each sequence in Test 8, whereas in Test 9, subjects recorded their responses online using AVRateNG,[12] running on another computer.

5.3.5 Subjective Evaluation Results

Outlier detection was performed to check the reliability of the subjects based on ITU-R Rec. BT.500-13 (BT.500-13 2012). In Test 8, no outlier was found. In Test 9, only one subject was detected as an outlier. MOS and associated 95% CI were computed for each of the test conditions in both tests.

[12] https://github.com/Telecommunication-Telemedia-Assessment/avrateNG

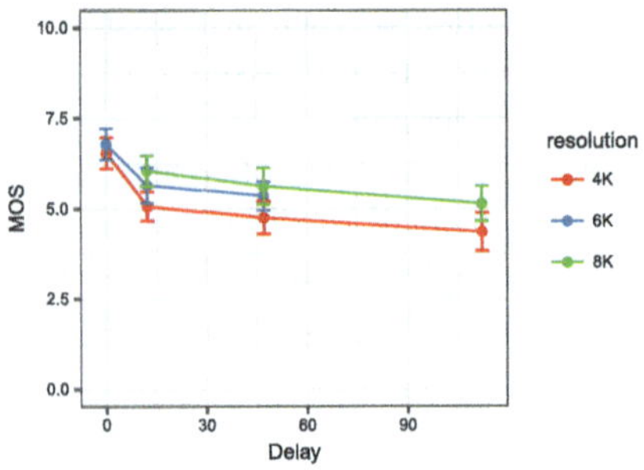

(a)

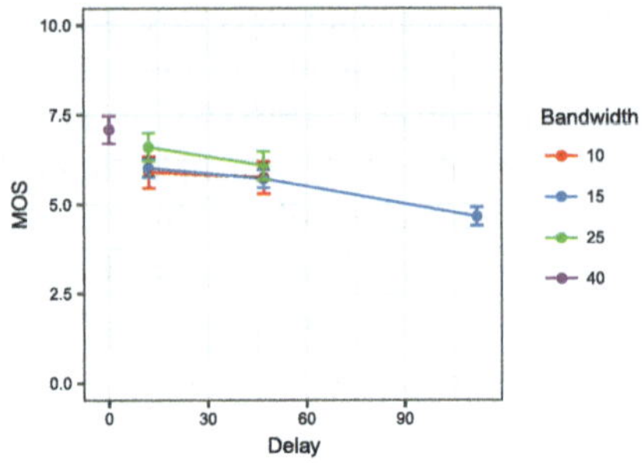

(b)

Fig. 5.6 MOS averaged over all contents with corresponding CIs for test 8 and 9 (Singla et al. 2019b). (**a**) Test 8. (**b**) Test 9

Table 5.5 ANOVA results for test 8

	DF	Sum Sq	Mean Sq	F	*p*
Delay	3	349.6	116.549	29.6311	<0.01
Resolution	2	98	48.981	12.4528	<0.01
Content	2	435.8	217.889	55.3957	<0.01
Delay:content	6	27.8	4.628	1.1765	0.3166
Delay:resolution	4	3.9	0.982	0.2496	0.9099
Resolution:content	4	7.7	1.913	0.4864	0.7457

Figure 5.6a shows average MOS over all video sequences for different delay values and resolutions. The selection of 3 different resolutions (4K, 6K, and 8K) was motivated by the resolution limitation of the HMD. One of the goals was to study whether subjects would be able to notice the difference in subjective video quality provided by the three different resolutions. Results (see Fig. 5.6a) show that subjects are able to detect the differences between the resolutions irrespective of the delay value. As expected, the subjective video quality degrades as the simulated network latency increases. In order to determine the impact of the experimental factors (resolutions, delay values, and video sequences) on the ratings, a within-subjects ANOVA was carried out on the individual ratings. Table 5.5 shows that delay, resolution, and content have statistically significant contributions with p-values <0.01.

Figure 5.6b shows the average MOS over all video sequences for different bandwidth and delay values. In Test 9, only a bandwidth-limited system for different values was tested. Results show that subjects perceive a difference in subjective video quality for different bandwidths at a given delay value. The subjective video quality declines with the increase in delay value. In order to find out the impact of the experimental factors (delay values, bandwidths, and video sequences) on the ratings, a within-subjects ANOVA was carried out on the individual ratings. Table 5.6 shows that bandwidth, delay, and content all have a statistically significant contribution, with p-values <0.01.

Table 5.6 ANOVA results for test 9

	DF	Sum Sq	Mean Sq	F	p
Delay	3	434.17	144.725	52.0232	<0.01
Bandwidth	2	27.73	13.863	4.9832	<0.01
Content	2	582.09	291.043	104.6191	<0.01
Delay:content	6	20.7	3.45	1.24	0.283127
Delay:Bandwidth	2	2.39	1.196	0.43	0.650614

Table 5.7 MCI, MOS Range, and MCI_{norm} for test 8 and 9 (Singla et al. 2019b)

	Test 8	Test 9
MCI	0.7635	0.686
MOS Range	4.107	4.541
MCI_{norm}	0.1859	0.1510

Furthermore, the results of the tile-based streaming with the playback of the full resolution were also compared in both tests as shown in Fig. 5.6a and b. Results show that users gave higher MOS to the anchor conditions over the tile-based streaming; but note that the anchor was local playback at a significantly higher bitrate without any delay. Interestingly, the difference between the 4K and 6K anchor conditions at 0 ms delay is quite minimal in the case of Test 8 (see Fig. 5.6a). A paired t-test was performed on the subjects' ratings for these two resolutions at 0 ms delay, showing that the difference is not statistically significant ($p = 0.2849$). To measure under which test conditions the results are more reliable, the statistical reliability was assessed based on Tominaga et al. (2010). MCI_{norm} was calculated as shown in Eq. 4.1, for both Test 8 and 9.

Table 5.7 shows the MOS Range, MCI, and MCI_{norm} for Test 8 and 9. Results indicate that evaluation results for the tile-based streaming solution evaluated under the bandwidth-limited scenario are more statistically reliable as compared to resolution and bandwidth-limited.

5.3.6 *Exploration Behavior*

For analyzing the head motion behavior of test subjects, all rotation angles around the single axes (pitch, yaw, and roll) were recorded, in synchronization with the played omnidirectional video. In the course of this section, it will be discussed whether there is an impact of the head rotation speed (in the yaw direction) on video quality scores. It was shown in Singla et al. (2017a) that users explore 360° video more in yaw than pitch direction. Equation 5.1 is used to calculate the speed in the yaw direction at a given time index n. S_n, Y_n and t_n are angular speed, yaw, and time values at time index n. The test 8 head rotation data is used for the analysis.

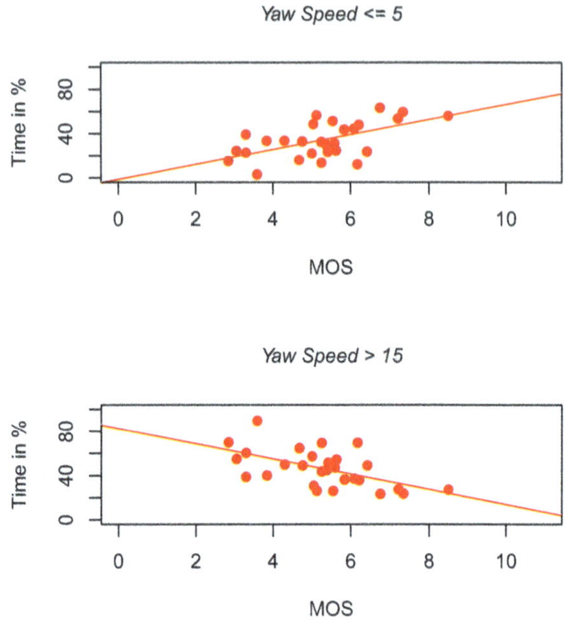

Fig. 5.7 Scatter plot between average time spent and averaged MOS for each subject for Test 8

$$S_n = \frac{abs(Y_n - Y_{n-1})}{t_n - t_{n-1}} \tag{5.1}$$

In viewport-adaptive streaming, if a person turns their head, depending on the delay, the viewport gets updated. It could be possible that if a person turns their head faster, then it may have an impact on video quality as in tile-based streaming; the viewport gets updated after a certain amount of delay. Therefore, it is important to investigate if subjects' head rotation speed has an impact on video quality. Here, two different values of speed are considered: 5°/s (low) and 15°/s (high).

Figure 5.7 (top) shows the scatter plot between average time spent for subject head rotation speed is less than or equal to 5 degrees/s in Yaw and MOS (averaged over all video sequences). If a person's head rotation speed is less than 5 degrees/s for most of the time, it is expected that the correlation between MOS and the time spent (in percentage) is positive (Pearson Correlation Coefficient = 0.57). This is due to the person not being able to notice distortions when they are moving their head.

Figure 5.7 (bottom) shows the scatter plot between the average time spent for subject head rotation speed is greater than 15 degrees/s in Yaw and MOS (averaged over all video sequences). If a person's head rotation speed is greater than 15 degrees/s for most of the time, it is expected that the correlation between MOS and the time spent (in percentage) is negative (Pearson Correlation Coefficient = −0.55). This is due to the person being able to notice distortions when they are moving their head.

Therefore, it could be concluded that if persons rotate their head slowly, then they may not be able to notice the distortions with smaller values of latency or the other way around, if persons rotate their head faster, then they are able to notice the distortions even with smaller values of latency.

5.4 Impact of Voting

In Gutierrez et al. (2022), subjective tests were carried out by ten laboratories in controlled environments. Each laboratory tested different test conditions. The author of the book in his lab carried out the test to check the influence of the two methods to collect the observers' ratings (i.e., through the controller and verbally). On the one side, the Unity-based application Miro360 (Pérez and Escobar 2019) was used (Test condition A), which allows the presentation and rating of the videos in the HMDs and the recording of the ratings and head rotation data. On the other side, the ratings can also be collected from the subjects verbally (Singla et al. 2017b) (Test condition B). In this case, the participant had to say the number of the rating aloud, and the experimenter noted it down. In both cases, the rating scales were displayed in the HMD after each test video, and the subjects were able to evaluate all the test videos without removing the HMD to rate.

The test 11 dataset is used and is described in Sect. 3.2.7.

5.4.1 Technical Settings and Test Method

The participants in the test were asked to freely watch and explore the contents and rate them in terms of audiovisual quality using the 5-point ACR scale. The test was conducted in a controlled environment, where the observers were seated in a swivel chair, so they could rotate freely to explore the 360° videos. The HTC Vive Pro (1440×1600 pixels per eye and 90 Hz) was used to watch the content.

The session structure is depicted in Fig. 5.8. Firstly, an introductory session was performed with the participants, where instructions for the test were provided, visual screening was performed, and training video samples were shown to appropriately adjust the HMD and familiarize them with the test methodology. Then, the participants evaluated the test stimuli corresponding to the first test condition (A)

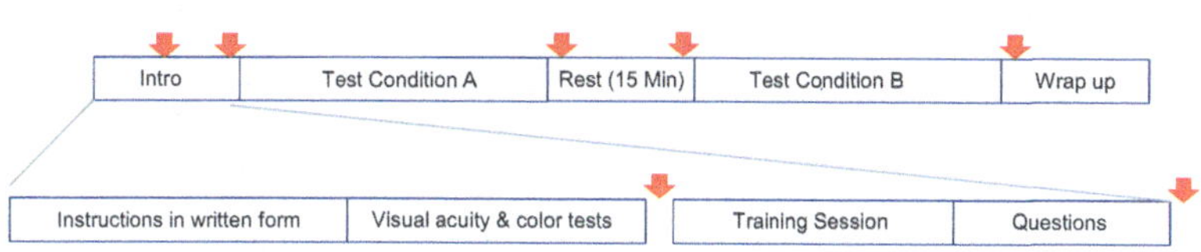

Fig. 5.8 Diagram of the structure of the test session for test 11 (Gutierrez et al. 2022)

and, after a break of 15 minutes, they evaluated the corresponding ones for the second test condition (B), according to be within-subject design.

As previously mentioned, the participants were asked to fill in questionnaires to evaluate simulator sickness (Kennedy et al. 1993), which was done various times during the test session. As depicted by the red arrows in Fig. 5.8, these questionnaires were filled in before the training session, after the training session and just before starting the evaluation of the first test condition (A), after the evaluation of the first test condition (A), after the rest and just before the evaluation of the second condition (B), and at the end of the test.

The whole test lasted less than 90 minutes, and the duration of each test session did not last more than 20 minutes, approximately. The non-uniform coding configurations using only 6×3 tiling patterns were used (See Fig. 3.14) to satisfy those time limits.

In total, 29 subjects participated in the subjective test. Out of 29, 14 were females and 15 were males. The range of subjects' age was between 20 to 37 with an average of 25.9 years.

5.4.2 Results

Outlier detection was performed to check the reliability of the subjects based on ITU-R Rec. BT.500-13 (BT.500-13 2012). No outlier was found in this test.

Figure 5.9 shows the MOS averaged over all video sequences with the corresponding 95% CI for both the rating methods. As expected, the subjective video quality degrades as the QP increases irrespective of the rating method. Furthermore, in this test 11, two different structures of tiles were used, smooth and abrupt transitions between adjacent tiles were considered as shown in Fig. 3.14 (Sect. 3.2.7). A t-test was performed on the averaged ratings (over all subjects) for those two configurations (abrupt and smooth), showing that the difference is not

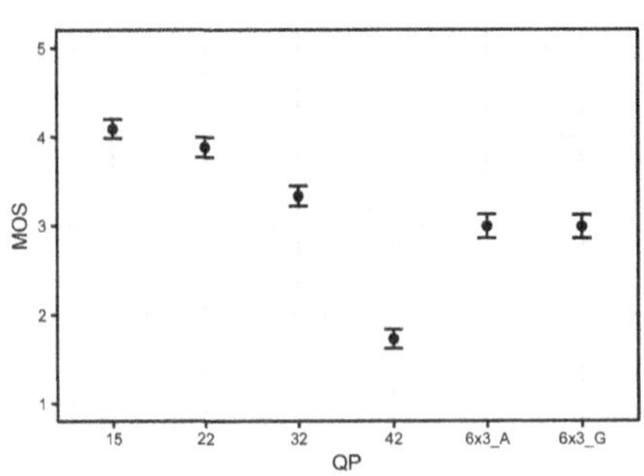

(a)

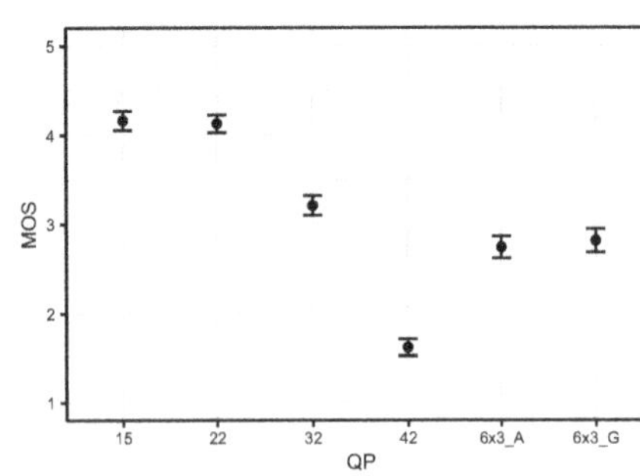

(b)

Fig. 5.9 MOS averaged over all contents with corresponding CIs for test 11. (**a**) Rating is provided using controller (Test Condition A). (**b**) Verbal rating (Test Condition B) significant ($p = 0.954$) when rated through the controller and $p = 0.426$ for the case of verbal rating.

The post-hoc Wilcoxon Signed-Rank[13] test showed no significantly different pairs among the 48 test videos compared. Therefore, both voting interfaces or verbal voting are recommended in the ITU-T (P.919 2020) for evaluations performed with 360° videos.

5.5 Prediction of Video Quality Scores for 360° Video

The objective of this section is to develop a model that can predict the 5-point MOS for a 360° video for a wide range of resolutions (FHD to 8K) at different bitrates. Multivariate linear regression is used to estimate the MOS for a given video using resolution (width×height), bitrate (in bits/s), and TI as variables. The TI value is calculated based on P.910 (2008), and the mean value of TI is used in the Eq. 5.2. The proposed model in this book follows a similar design philosophy as the Hybrid Mode 0 model proposed by Fremerey et al. (2020).

For the development of the mathematical model, the test 1, 2, and 6 data was used. The dataset was equally divided into two subsets, one for training and one for evaluation without any overlap between these two subsets. The video #1, #2, and #3 were used for training from test 1 and 2. From test 6, the video #1, #4, and #5 were used for training. For the evaluation of model, the video #4, #5, and #6 were used from test 1 and 2. From test 6, the video #2 and #3 were used for the evaluation. The training of the model was carried out in R using LmFit. The coefficients for the model are summarized in Table 5.8. Equation 5.2 is used to obtain predicted MOS (MOS_P) and Eq. 5.3 is used to limit the predicted value of MOS in the range of $[1, 5]$.

$$MOS_P = a \cdot \log(bitrate) + b \cdot \log(resolution) + c \cdot \log(TI) + d \tag{5.2}$$

$$MOS_{Final} = \max(1, \min(5, MOS_P)) \tag{5.3}$$

Table 5.8 Proposed model coefficients

a	b	c	d
1.78	0.31	−0.821	−10.21

[13] The reason for using mixed models analysis is to compare means which would technically not allow using classical linear regression. Furthermore, the interest is in the general differences among the test conditions (Gutierrez et al. 2022).

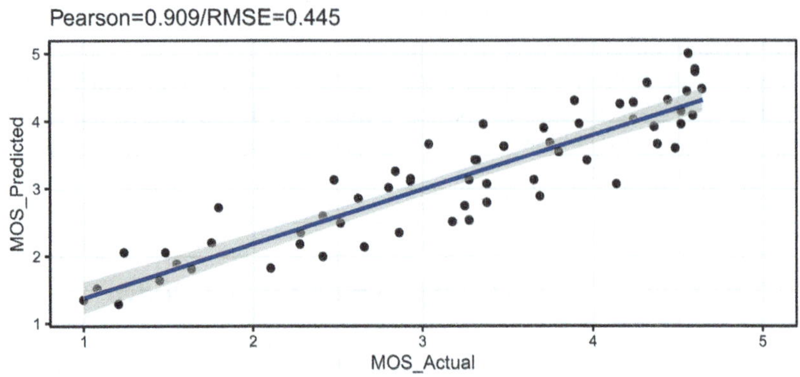

Fig. 5.10 Scatter plot for the proposed model

To evaluate the performance of the model, the Pearson Correlation Coefficient (PCC) is calculated between ground-truth (referred to as actual) data and the estimated (referred to as predicted) scores. Furthermore, the value of Root Mean Square Error (RMSE) is also calculated. The model shows a high prediction performance on the validation dataset, with a Pearson correlation of 0.909 and RMSE of 0.445. A scatter plot is plotted between the predicted and actual MOS values as shown in Fig. 5.10.

5.6 Recommendations

Based on the studies reviewed in the Chap. 2 and based on the experimental results in Chaps. 4 and 5, recommendations/guidelines can be derived for conducting tests for media quality assessment of 360° videos.

Before starting the test, the test participants should be screened for correct visual acuity using Snellen charts (20/25) and for color vision using Ishihara charts, as described in ITU-R Rec. BT.500-13. During post-test subject screening, outlier detection can be performed based on the PCC as described in Tan et al. (2014) or based on ITU-R Rec. BT.500-13.

For playing out 360° videos, a player must be used that provides smooth playback for 360° videos without stuttering, such as Whirligig (Hofmeyer et al. 2019). It was shown that for reducing the quality-impact due to the mismatch between video frame rate and HMD refresh rate, and hence to avoid stuttering, the video frame rate should be adapted to the HMD's refresh rate (Hofmeyer et al. 2019). For interpolating the sequences from the given frame rate to the 90 Hz used by most HMDs, FFmpeg MCI and butterflow algorithms are suggested, depending on the amount of motion in the video (Hofmeyer et al. 2019). To provide at least a reasonable level of perceived quality, video resolution should be at least 4K (Singla et al. 2017b, 2018; Tran et al. 2017).

For the evaluation of short video sequences ($\leq$20 s), for a more fine-grained video quality assessment, a double-stimulus test method should be used, such as DSIS. In case that an absolute rating is sought, an adapted single-stimulus method such as the proposed M–ACR can be used, where the test sequences are repeated twice, so that test participants have enough time to explore the omnidirectional videos and assess their quality.

Using longer sequences ($\geq$20 s) enables a more realistic and holistic viewing experience, since subjects can better explore and adjust to the current content. In this case, a double-stimulus method is not suitable. Instead, a single-stimulus test method such as ACR can well be used to avoid a longer duration of the test.

For the collection of ratings, either test participants could record their responses using a controller (Gutierrez et al. 2022) or report the rating number aloud to an experimenter present in the room, who notes down the rating of the video (Singla et al. 2017b). Both approaches enable subjects to continuously wear the HMD, avoiding possibly detrimental effects of taking on and off the HMD when collecting ratings outside, e.g., on a separate screen, or on a paper. It was shown that particular care must be taken when different subjective tests are to be compared, where contextual effects such as outside-the-lab temperatures may have a stronger effect on the results (Singla et al. 2019a). These considerations as well as further methodological aspects proposed during the work have been integrated into the new standard ITU-T Rec. P.919 (P.919 2020).

References

J.G. Beerends, F.E. De Caluwe, The influence of video quality on perceived audio quality and vice versa. J. Audio Eng. Soc. **47**(5), 355–362 (1999). http://www.aes.org/e-lib/browse.cfm?elib=12105

ITU-R BT.500-13, *Recommendation ITU-R BT.500-13: methodology for the subjective assessment of the quality of television pictures* (2012)

S. Fremerey et al., Subjective test dataset and meta-data-based models for 360° streaming video quality, in *2020 IEEE 22nd International Workshop on Multimedia Signal Processing (MMSP)* (2020), pp. 1–6

J. Gutierrez et al., Subjective evaluation of visual quality and simulator sickness of short 360° videos: ITU-T Rec. P.919. IEEE Trans. Multimedia **24**, 3087–3100 (2022). https://doi.org/10.1109/TMM.2021.3093717

F. Hofmeyer et al., Impacts of internal HMD playback processing on subjective quality perception. Electron. Imaging **2019**(12), 219–1 (2019)

S. Ishihara, *The Series of Plates Designed as a test for Colour Deficiency*. http://www.dfis.ubi.pt/~hgil/P.V.2/Ishihara/Ishihara.24.Plate.TEST.Book.pdf (16 May 2018) (1997)

R.S. Kennedy et al., Simulator sickness questionnaire: an enhanced method for quantifying simulator sickness. Int. J. Aviation Psychol. **3**(3), 203–220 (1993)

ITU-T Rec. P.910, *Recommendation ITU-T P.910: subjective video quality assessment methods for multimedia applications* (2008)

ITU-T P.915, *Subjective assessment methods for 3D video, P.915* (2016)

ITU-T P.919, *Subjective test methodologies for 360° video on head-mounted displays*. Recommendation P.919 (2020)

P. Pérez, J. Escobar, MIRO360: a tool for subjective assessment of 360 degree video for ITU-T P.360-VR, in *2019 Eleventh International Conference on Quality of Multimedia Experience (QoMEX)*, June 2019, pp. 1–3. https://doi.org/10.1109/QoMEX.2019.8743216

A. Singla et al., AhG8: measurement of user exploration behavior for omnidirectional (360-degree) videos with a head mounted display, in *Joint Video Exploration Team of ITU-T SG16 WP3 and ISO/IEC JTC1/SC29/WG11, JVET-H0050, 8th Meeting*, Oct 2017a

A. Singla et al., Comparison of subjective quality evaluation for HEVC encoded omnidirectional videos at different bit-rates for UHD and FHD resolution, in *Proceedings of the on Thematic Workshops of ACM Multimedia*, Mountain View, 2017b. ISBN: 978-1-4503-5416-5. https://doi.org/10.1145/3126686.3126768

A. Singla et al., Measuring and comparing QoE and simulator sickness of omnidirectional videos in different head mounted displays, in *9th International Conference on QoMEX*, May 2017c, pp. 1–6. https://doi.org/10.1109/QoMEX.2017.7965658

A. Singla, W. Robitza, A. Raake, Comparison of subjective quality evaluation methods for omnidirectional videos with DSIS and Modified ACR. Electron. Imaging **2018**, 1–6 (2018). https://doi.org/10.2352/ISSN.2470-1173.2018.14.HVEI-525

A. Singla, W. Robitza, A. Raake, Comparison of subjective quality test methods for omnidirectional video quality evaluation, in *2019 IEEE 21st International Workshop on Multimedia Signal Processing (MMSP)*, Sept 2019a, pp. 1–6

A. Singla et al., Subjective quality evaluation of tile-based streaming for omnidirectional videos, in *Proceedings of the 10th ACM Multimedia Systems Conference*, MMSys'19, Amherst (ACM, 2019b), pp. 232–242. ISBN: 978-1-4503-6297-9. https://doi.org/10.1145/3304109.3306218

T.K. Tan et al., HEVC subjective video quality test results, in *IBC2014 Conference, 2014*, Amsterdam, 11–15 Sept 2014 (Institution of Engineering and Technology, UK, 2014). ISBN: 978-1-84919-927-8. https://doi.org/10.1049/ib.2014.0029

A. Tirumala et al., *Iperf: The TCP/UDP Bandwidth Measurement Tool*. https://iperf.fr/ (20 June 2018) (2005)

T. Tominaga et al., Performance comparisons of subjective quality assessment methods for mobile video, in *2010 Second International Workshop on QoMEX*, June 2010, pp. 82–87. https://doi.org/10.1109/QOMEX.2010.5517948

H.T.T. Tran et al., A subjective study on QoE of 360 video for VR communication, in *IEEE 19th International Workshop on MMSP* (2017), pp. 1–6

Chapter 6
Simulator Sickness Assessment for 360° Videos

6.1 Impact of Technical Factors

There are a number of video-related technical factors such as resolution, bitrate, motion, audio, frame rate, and projection scheme that can contribute to simulator sickness scores. This section discusses the impact of resolution and bitrate on the scores of simulator sickness.

6.1.1 Resolution

The test 1 dataset is used to assess the impact of resolution on simulator sickness scores of two different HMDs, namely HTC Vive and Oculus Rift, and is described in Sect. 3.2.1. The technical setup and test method are described in Sect. 5.1.

For this experiment, the content was categorized based on informal contemplation. The source videos were selected to include some sequences expected to make the user uncomfortable, e.g., due to high amounts of motion. It can be seen from Fig. 6.1 that contents 1, 2, and 4 lead to high scores of Nausea (N), Oculomotor (O), Disorientation (D), and TotalScore (TS). These contents also have a higher value of TI (see Fig. 3.3). Contents 3, 5, and 6 lead to low scores, irrespective of the device and the resolution. It appears to be from the results that users' tend to be more prone to simulator sickness in those video sequences which have higher motion and less prone to simulator sickness in those sequences that have less motion. Similar results were found by Tran et al. in (2017), who reported that users feel uncomfortable in those contents that have higher motion content in them.

In order to find out whether users are more prone to simulator sickness in 4K or FHD resolutions, the difference between the sickness scores of 4K and FHD resolutions is calculated for both HMD devices. Figure 6.2 shows the difference in

A. Singla, *Assessment of Visual Quality and Simulator Sickness for Omnidirectional Videos*, T-Labs Series in Telecommunication Services,
https://doi.org/10.1007/978-3-031-49988-3_6

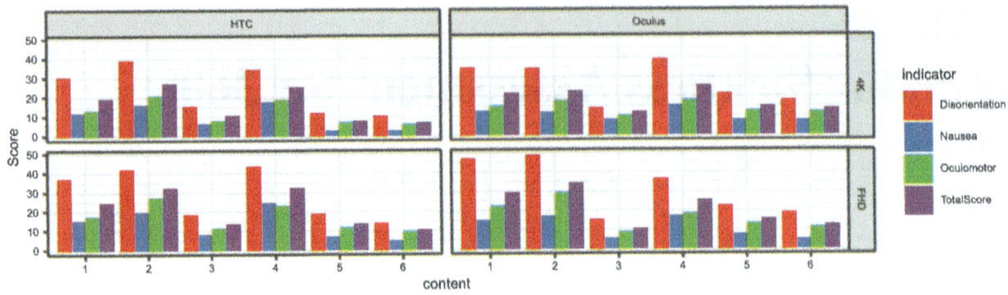

Fig. 6.1 Simulator sickness scores for all devices and resolutions for test 1 (Singla et al., 2017b)

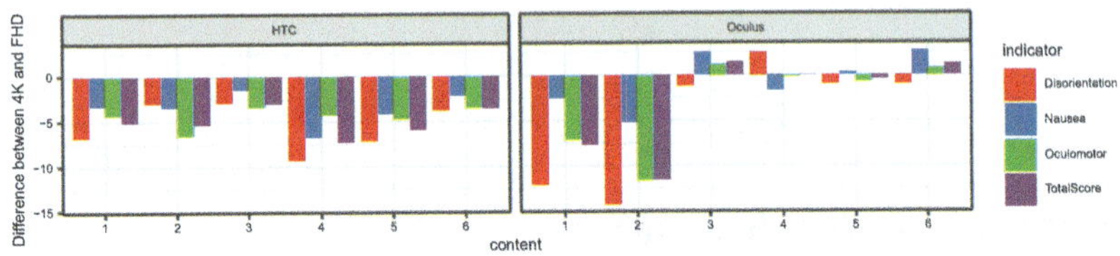

Fig. 6.2 Difference in Simulator sickness scores depending on resolution for test 1 (Singla et al., 2017b)

the simulator sickness scores in 4K or FHD resolutions for the HTC Vive and Oculus Rift. Positive bars mean that users have experienced more simulator sickness in 4K resolution as compared to FHD. It is clear that users felt more prone to simulator sickness in FHD resolution for HTC Vive. For Oculus Rift, users have experienced low simulator sickness for contents 1 and 2 in 4K. Contents 3 and 6 lead to slightly lower simulator sickness scores for FHD resolution as compared to 4K resolution.

For some of the sequences, users have experienced more sickness in HTC Vive than Rift. Therefore, based on the results, it could not be deduced that users are less prone to simulator sickness in HTC Vive than Oculus Rift while watching videos. Similar results were found by Yildirim (2019) where they compared HTC Vive with Oculus Rift CV 1. They found that the level of cybersickness induced by both HMDs is comparable. Hence, it can be concluded that lenses and rendering algorithms between these HMDs do not have an impact on the ratings of simulator sickness. Please note that HTC Vive and Oculus Rift have similar technical specifications (resolution, refresh rate, and FOV).

Finally, an ANOVA was carried out on the TS scores with devices, resolution, and content as independent variables. Results indicate that content and resolution have a significant impact ($p < 0.01$), whereas the device does not.

6.1.2 *Bitrate*

The test 10 dataset (see Sect. 3.2.6) is used to assess the impact of bitrate on comfort scores.

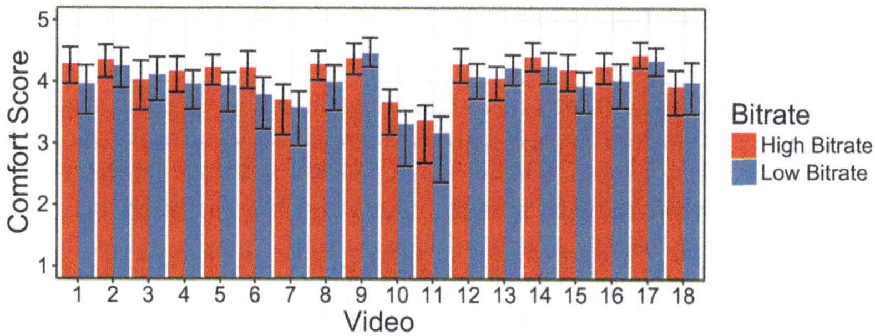

Fig. 6.3 Comfort scores for all sequences with corresponding CIs for different bitrates for test 10

In this study, 360° videos were presented to participants in full resolution. After each video, participants were asked to rate video quality (to align this research with related 360°-video research) followed by a comfort question that reports their viewing comfort level. To enable short-term ratings in relation to simulator sickness, comfort ratings were used (see B.2 for the exact question).

The 360° videos were viewed on the HTC Vive Pro using the *Whirligig* player. The FOV was set to 360 and projection type to *Barrel*. The resolution of the HMD is 2880 × 1600 and its FOV is 110° horizontally.

The ACR (ITU-T Rec. P.910 2008) test method was selected to evaluate audiovisual quality. For comfort, a 5-point scale was used according to ITU-R BT.2021 (see B.2). 28 participants (14 female, 14 male) participated in the test. Visual acuity (20/25, Snellen) and color blindness (Ishihara) were tested. The test was preceded by a training session for test-subject familiarization. Each subject had to rate 36 PVSs (18 sequences in 2 bitrates) over three test sessions. First, the participant had to watch the video for 60 s in an HMD and had 6 s each to provide the quality and comfort scores. Before and after each test session, participants filled the full SSQ (Kennedy et al., 1993).

A scale inversion was carried out by subtracting the comfort scores from the value 6. Figure 6.3 shows the resulting inverted comfort scores for all video sequences for different bitrates averaged over all participants. It is evident from the plot that the participants felt more comfortable in watching 360° videos at higher bitrate for all sequences except for sequences #3, #9, #13, and #18. It is interesting to note that the sequences (#10 and #11) that have a higher amount of motion (see Fig. 3.11, where it can be seen that these sequences have a higher value of TI) provide the lowest level of comfort to the participants irrespective of the bitrate. As such, it can be said that at higher bitrates users feel more comfortable than with lower bitrates at the same resolution.

To find out the impact of video bitrate on discomfort, a repeated-measures ANOVA was carried out on the individual ratings, indicating a statistically significant contribution ($p < 0.01$).

6.2 Exploration Behavior

In Singla et al. (2017a), it was investigated by the author to what extent users are exploring 360° videos in yaw and pitch directions. The results showed that most of the subjects have a relative range of 40° and 110° in the pitch and yaw direction, respectively. The possible reason could be that users may not feel comfortable moving their head upwards/downwards or maybe it was due to the contents not having pitch-relevant information. Furthermore, it was observed that in the sequences which have camera motion, most of the time users tend to be stationary as compared to those sequences where the camera is stationary.

Therefore, for the analysis conducted in this section, exploration behavior only in the yaw direction will be considered. Therefore, it is desirable to investigate if high/low exploration behavior results in high/low simulator sickness scores. Hence, the data from test 1 is used (Singla et al., 2017b) for the analysis. Due to technical reasons, the head rotation data of four subjects could not be recorded. Hence, the head rotation data only for 24 subjects were considered. The yaw values were measured after every 0.15 s, therefore, for a 60 s sequence approx. 400 values were recorded.

Here, the main aim was to find if subjects who are more prone to the symptoms of simulator sickness have less exploration behavior or vice versa. First, the distance covered by each participant for each PVS was calculated. For that, the absolute difference between the consecutive values in the yaw direction was obtained and summed over all the time indexes. Then, the average yaw distance and Total Score (Simulator Sickness) across all PVSs for each subject was calculated. Figure 6.4 shows the scatter plot between average yaw distance and total score for all test subjects. Blue lines indicate the bifurcation of exploration distance and Total Score into high and low. From the results, it is not clear whether subjects who have high sickness scores have explored less or vice versa.

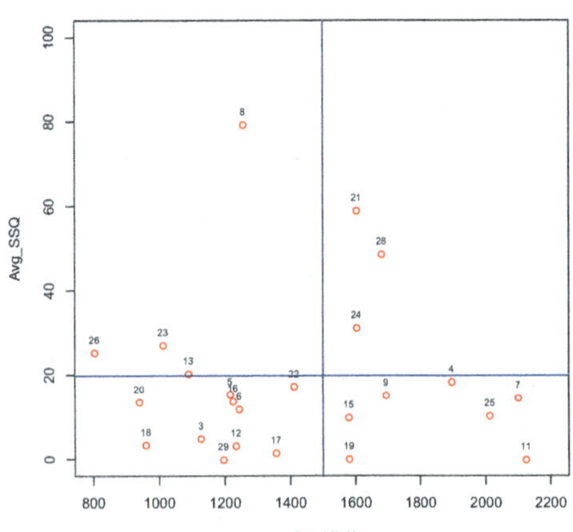

Fig. 6.4 Scatter plot comparing average yaw distance and TotalScore for each subject for test 1

Based on the above result, it cannot be concluded if there is a link between simulator sickness and head movements.

6.3 Tile-Based Streaming

While assessing the quality of tile-based 360° video streaming in an HMD, users may experience symptoms of simulator sickness as well. Here, what is being shown on the screen actually depends on the previous and current head position, depending on the applied algorithms for tile-based streaming and content pre-fetching. Consequently, cybersickness may differently depend on exploration behavior in this case than for traditional "full frame streaming." Therefore, evaluating simulator sickness in the context of the proposed test design is essential. In neither of the two tests (Test 8 and 9) did any of the subjects leave the experiment before it ended due to simulator sickness.

The tests 8 and 9 dataset are described in Sect. 3.2.5. The technical setup and test method are described in Sect. 5.3.

6.3.1 Influence of Content on SSQ Scores

In tests 8 and 9, for evaluating the impact of the video sequence on the simulator sickness scores, a short simulator sickness questionnaire (Short SSQ) (Tran et al., 2017) was used. In the Short SSQ (See B.2), subjects provided their rating on a 5-point scale (Very dizzy: 5, dizzy: 4, slightly dizzy: 3, not dizzy: 2, absolutely not dizzy: 1). Subjects completed the Short SSQ after every test video. Figure 6.5a and b show that video #2 (Balboa) has the highest scores of simulator sickness, irrespective of the test. In video #2, a camera is mounted on a driving car. This

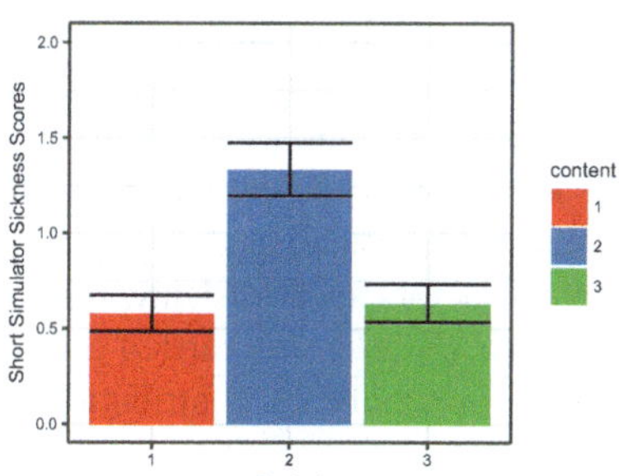

(a)

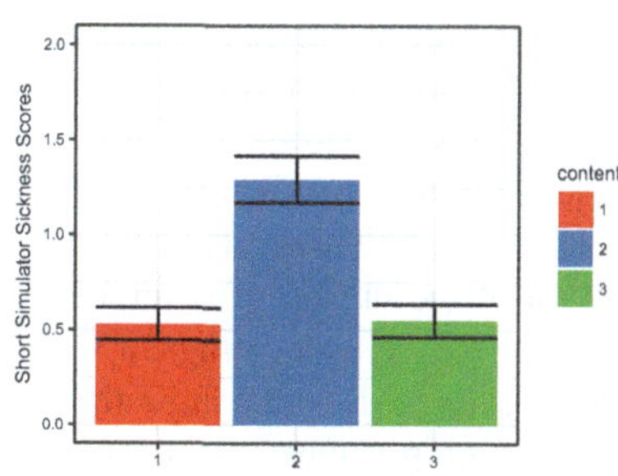

(b)

Fig. 6.5 Short simulator sickness Scores for all contents averaged across all subjects with corresponding CIs for test 8 and 9 (Singla et al., 2019). (**a**) Test 8. (**b**) Test 9

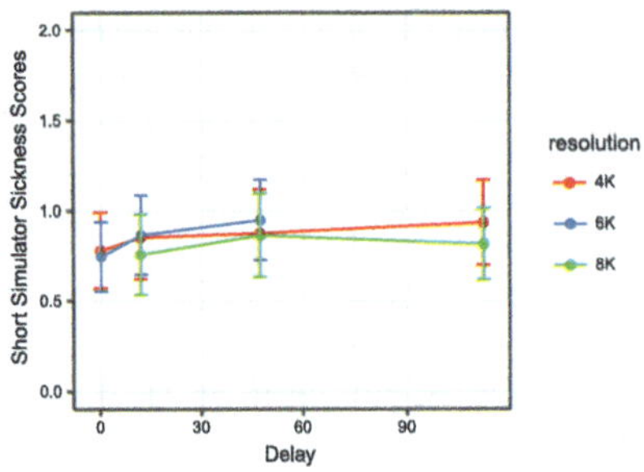

(a)

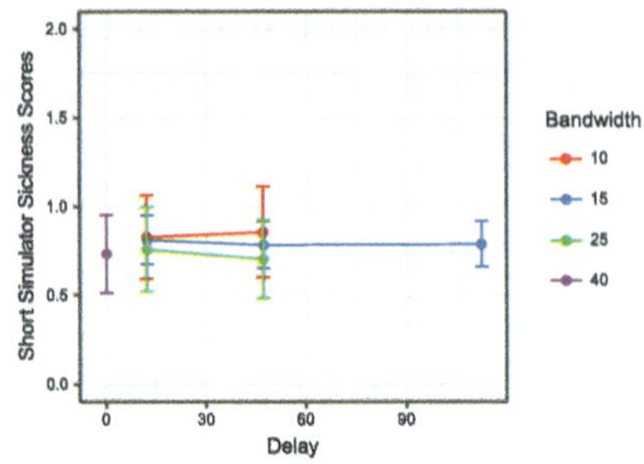

(b)

Fig. 6.6 Short simulator sickness Scores averaged across all subjects and contents for different delay values with corresponding CIs for test 8 and 9 (Singla et al., 2019). (**a**) Test 8. (**b**) Test 9

video also has the highest value for TI, as shown in Fig. 3.9. Videos # 1 and #3 have almost similar scores for the simulator sickness in both tests. Also, these two video sequences have similar values of TI as shown in Fig. 3.9.

Based on the results obtained in Sects. 6.1 and 6.3.1, it can be deduced that users are more prone to simulator sickness when the video has a high motion in it.

6.3.2 Influence of Network Delay on SSQ Scores

In this section, it is investigated if the network delay has an impact on the Short SSQ scores. Figure 6.6a and b show the Short SSQ scores averaged over all the video sequences and subjects for different delay values. From the results, it is clear that delay (motion-to-high-resolution latency, see Sect. 2.4) does not have an effect on the simulator sickness scores for either test. This may be explained by the fact that even if the low-resolution background is visible, it always moves consistently with head motion. In other words, even when motion-to-high-resolution latency increases, motion-to-photon latency remains at the same low value. It is interesting to note that users felt more prone to simulator sickness while testing the video quality under bandwidth- and resolution-limited conditions (Test 8) than in the bandwidth-limited test (Test 9), see Fig. 6.6a and b. Any concrete understanding of this behavior under the current experiment could not be established. For a statistical evaluation of the impact of several experimental factors such as delay, resolution, and content on the user ratings, a within-subjects ANOVA was carried out on the individual ratings. Results show that content has a significant contribution to the SSQ score with p-values <0.01 for both of the tests.

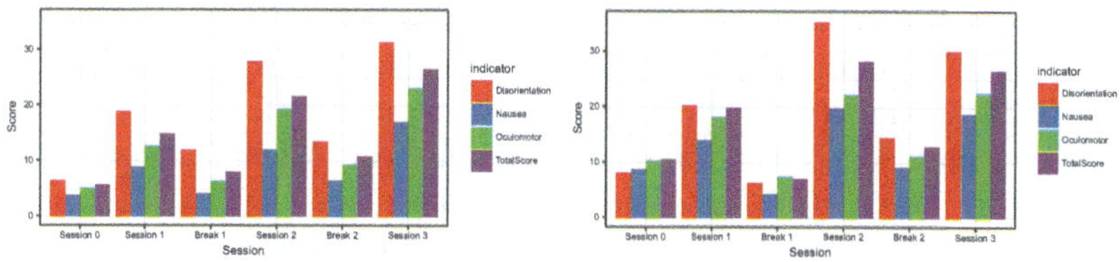

(a) (b)

Fig. 6.7 Simulator sickness scores for all test sessions for test 8 and 9 (Singla et al., 2019). (**a**) Test 8. (**b**) Test 9

6.3.3 Influence of Test Duration on SSQ Scores

The duration of each test session for Tests 8 and 9 was 10 and 14 min, respectively. In the following, it will be studied whether the test session and overall test duration have an impact on simulator sickness scores. Figure 6.7 shows the average simulator sickness scores over all test subjects for Tests 8 and 9. To investigate whether breaks during the tests lead to a decrease in the simulator sickness scores, subjects were asked to fill in the long simulator sickness questionnaire two times: (1) just after the test session and (2) after the break, just before starting the next test session. For both tests, the SSQ scores were also assessed for session #0, which were collected just before the training session and after the pre-screening.

Results suggest that simulator scores are increasing with session time for both tests. Also, it can be seen that breaks help in reducing the simulator sickness, which highlights the significance of breaks during tests like these. From Fig. 6.7b, it is clear that the SSQ scores for Test 9 are higher as compared to the scores of Test 8 except for session #3. The possible reason could be after session #2, some of the users have adapted to the viewing environment and hence, the severity of the symptoms of simulator sickness is reduced. In Test 9, users evaluated 14 PVSs as compared to 10 PVSs in Test 8. The correspondingly higher simulator sickness scores clearly underline that the duration of the test session has an impact on the SSQ scores.

6.4 Toward Simulator Sickness Modeling

Predicting simulator sickness while watching 360° videos is of relevance to ensure viewing comfort and safety. This way problems may be anticipated even before users view certain contents, or during a viewing session, even before cybersickness symptoms start hitting users more badly. The symptoms of simulator sickness may not be eliminated completely by making advancements in software and hardware (such as tracking, resolution of the device and displayed video, higher refresh rate, improved production quality with regard to, e.g., stitching, etc.).

In 2019, Kim et al. proposed a VR sickness assessment CNN (VRSA) that takes into account motion patterns of VR videos and predicts a sickness score for a test video. In 2018, Kim et al. proposed a framework called "VR sickness predictor" (VRSP) that considers perceptual motion and statistical content features to predict sickness scores on an individual-subject level and per video. In 2018, Padmanaban et al. designed a sickness predictor that estimates the nauseogenicity (i.e., tendency to introduce nausea) of 360° stereoscopic videos.

All the aforementioned studies predict sickness for specific videos or VR scenes, or at an individual-subject level. None of the reviewed studies addresses sickness more holistically with respect to its evolution over time and in light of possible pauses or other actions during a viewing session.

The contributions of the section are as follows: (1) First, introducing a novel general concept of a simulator sickness model applicable to complete viewing sessions, (2) the prediction of simulator sickness over time, taking breaks into account.

6.4.1 SiSiMo Concept

In the running work on developing SiSiMo, it is assumed that simulator sickness results from a temporal succession of different phases that may positively or negatively impact simulator sickness and result in a certain level of overall comfort. Each phase is addressed by a specific module within SiSiMo (cf. Fig. 6.8). Here different phases are distinguished, namely, of *viewing* (module M_V, may include audio), where, e.g., a high-motion video may lead to an increase of simulator sickness, *pausing* (M_P), assumed to lead to a decrease in simulator sickness, and *rating* (M_R), reflecting that respective research typically involves ratings of simulator sickness or additional aspects such as video quality. It is noted that the latter will also include aspects of the actual rating procedure used. For the viewing module M_V, approaches as discussed in Kim et al. (2019) and Padmanaban et al. (2018) are being considered. In addition, the model concept includes inter- and intra-phase *adaptation* ("A") between instances of the aforementioned phases *viewing*, *pausing*, and *rating*, comprising aspects such as taking on and off the HMD, scene changes or between-video switches, and switching to carrying out rating tasks within the HMD, i.e., with the scale shown on its screen, or on paper without wearing the HMD.

The simulator sickness contributions of the different components are integrated into the overall simulator sickness score in model component "I" (cf. Fig. 6.8). To develop this rather complex model, a series of subjective tests needs to be conducted to analyze each factor specifically and in terms of interactions. As a first step, the evolution of simulator sickness over time is considered.

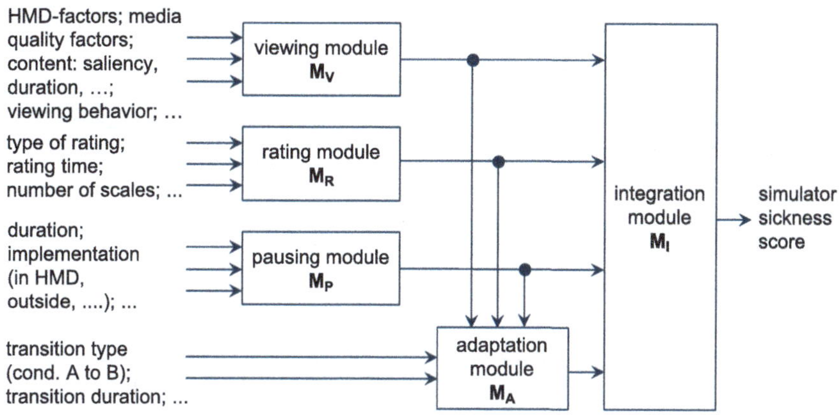

Fig. 6.8 General concept of SiSiMo model (Raake et al., 2020)

6.4.2 *Technical Setup and Equipment*

The test 10 dataset (see Sect. 3.2.6) is used to develop the model.

The 360° videos were viewed on the HTC Vive Pro using the *Whirligig* player. The Field of View (FOV) was set to 360 and the projection type to *Barrel*. The resolution of the HMD is 2880×1600 and its FOV is 110° horizontally. The HMD was connected to a desktop PC with a current CPU (Intel i7, 6700), NVIDIA GTX1080 graphics card and SSD. To record the viewing direction (yaw, pitch, and roll) of the participants, an open-source framework[1] was used.

6.4.3 *Test Method*

The Absolute Category Rating (ACR) (ITU-T Rec. P.910) test method was selected to evaluate the audiovisual quality. For comfort, a 5-point scale was used according to ITU-R BT.2021. 28 participants (14 female, 14 male) participated in the test with an average age of 26.1 years. Visual acuity (20/25, Snellen) and color blindness (Ishihara) were tested. The test was preceded by a training session for test-subject familiarization. Each subject had to rate 36 Processed Video Sequences (PVSs, 18 sequences in 2 bitrates) over three test sessions. First, the participant had to watch the video for 60 s in an HMD and had 6 s each to provide the quality and comfort scores. Before and after each test session, participants filled in the full Simulator Sickness Questionnaire (SSQ) (Kennedy et al., 1993).

[1] https://github.com/acmmmsys/2018-AVTrack360

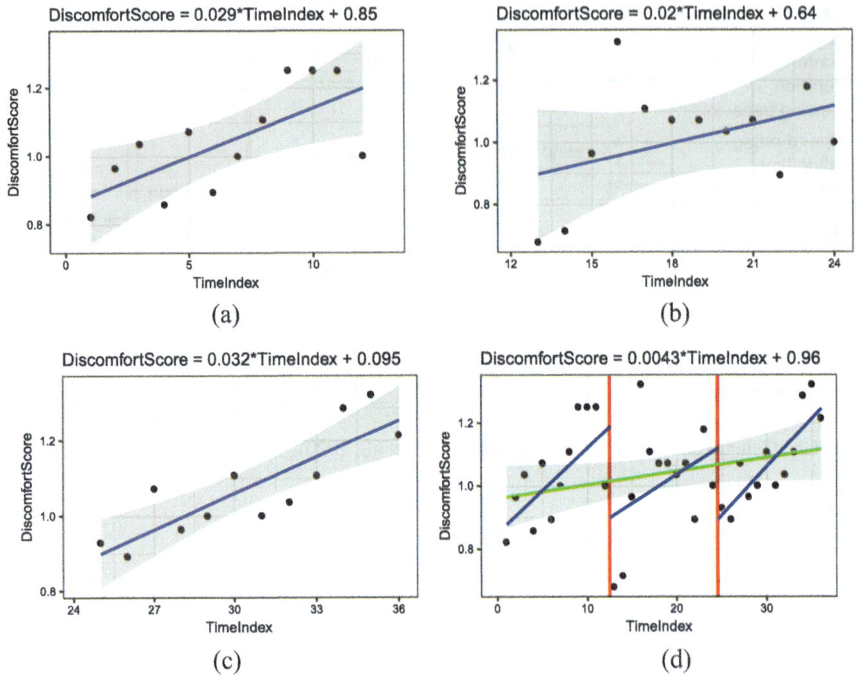

Fig. 6.9 Scatter plot and linear regression lines indicating correlation between time index and discomfort scores (Raake et al., 2020). (**a**) Session 1. (**b**) Session 2. (**c**) Session 3. (**d**) All Sessions

6.4.4 Results and Temporal Evolution Model

Outlier detection to check the reliability of test participants was performed on quality ratings based on ITU-R Rec. BT.500-13, and no outliers were found.

The resulting current discomfort score (as indicator of simulator sickness) is assumed to primarily depend on the specific video sequence s_k and time index t_i of viewing it. To address temporal evolution in this section, the discomfort scores for sequences with the same time index t_i were averaged, and are shown in Fig. 6.9. Results for the three sessions are depicted in Fig. 6.9a–c. Figure 6.9d shows the results for the entire test. Here, the red lines indicate the point in time where the subjects were taking a break between two test sessions.

As starting point to model development a linear regression was used with regard to viewing time t_i. Here it is assumed that discomfort increases over time and is gradually reduced during breaks between sessions. Linear regression was hence applied separately on each test session; see plots (a)–(c). To evaluate the performance of the proposed model, the PCC between subjective scores (Comfort Scores) and SiSiMo estimates was calculated. The resulting PCC values for sessions 1, 2, and 3 are 0.68, 0.40, and 0.84, respectively. The overall (green) and per-session regression lines (blue) in (d) indicate a saw-tooth like pattern, with increasing discomfort per session as well as a slight increase across sessions, in spite of the breaks. An analysis of the relation between discomfort and the source video sequences revealed the known link between content and simulator sickness also discussed in Kim et al. (2018, 2019) and Padmanaban et al. (2018).

6.5 Conclusions

In this chapter, the influence of different factors such as video resolution, bitrate, session duration, and latency on cybersickness was investigated. From the results, it can be derived that resolution and bitrate of the video should be high to have less severity of the symptoms of simulator sickness. Users are more comfortable watching a higher resolution/bitrate video than in a lower resolution/bitrate. Furthermore, videos that have higher motion content should be avoided to have a more pleasant experience for users in HMDs.

For viewport-adaptive streaming, simulator sickness was assessed after every test condition and test session. For the specific tile-based streaming technology used, the results show that the effect of network delay on the simulator sickness scores cannot be seen. The test results indicate that simulator scores increase with session time and breaks help in reducing the simulator sickness in both tests.

To predict and thus enable alleviation of simulator sickness, a conceptual Simulator Sickness Model *SiSiMo* is proposed to predict sickness over longer viewing sessions. The first results show that temporal evolution of discomfort, as a simplified measure of simulator sickness, can be estimated using linear regression over the viewing-time index for 360° videos. The performance of the model is promising; however, the prediction accuracy is moderate for some of the sessions (1 and 3). Here additional consideration of the actual video viewed and the individual proneness to simulator sickness will be included for future improvements.

References

R.S. Kennedy et al., Simulator sickness questionnaire: an enhanced method for quantifying simulator sickness. Int. J. Aviat. Psychol. **3**(3), 203–220 (1993)

J. Kim et al., Virtual reality sickness predictor: analysis of visual-vestibular conflict and VR contents, in *2018 Tenth International Conference on Quality of Multimedia Experience (QoMEX)* (2018), pp. 1–6. https://doi.org/10.1109/QoMEX.2018.8463413

H.G. Kim et al., VRSA net: VR sickness assessment considering exceptional motion for 360-degree VR video. IEEE Trans. Image Process. **28**(4), 1646–1660 (2019). https://doi.org/10.1109/TIP.2018.2880509

ITU-T Rec. P.910, *Recommendation ITU-T P.910: subjective video quality assessment methods for multimedia applications* (2008)

N. Padmanaban et al., Towards a machine-learning approach for sickness prediction in 360-degree stereoscopic videos. IEEE Trans. Vis. Comput. Graph. **24**, 1594–1603 (2018)

A. Raake et al., SiSiMo: towards simulator sickness modeling for 360-degree videos viewed with an HMD, in *2020 IEEE Conference on Virtual Reality and 3D User Interfaces Abstracts and Workshops (VRW)* (2020), pp. 582–583

A. Singla et al., AhG8: measurement of user exploration behavior for omnidirectional (360-degree) videos with a head mounted display, in *Joint Video Exploration Team of ITU-T SG16 WP3 and ISO/IEC JTC1/SC29/WG11, JVET-H0050, 8th Meeting* (2017a)

A. Singla et al., Measuring and comparing QoE and simulator sickness of omnidirectional videos in different head mounted displays, in *9th International Conference on QoMEX* (2017b), pp. 1–6. https://doi.org/10.1109/QoMEX.2017.7965658

A. Singla et al., Subjective quality evaluation of tile-based streaming for omnidirectional videos, in *Proceedings of the 10th ACM Multimedia Systems Conference*, MMSys'19, Amherst (ACM, 2019), pp. 232–242. ISBN: 978-1-4503-6297-9. https://doi.org/10.1145/3304109.3306218

H.T.T. Tran et al., A subjective study on QoE of 360 video for VR communication, in *IEEE 19th International Workshop on MMSP* (2017), pp. 1–6

C. Yildirim, Don't make me sick: investigating the incidence of cybersickness in commercial virtual reality headsets. Virtual Reality **24**, 231–239 (2019)

Chapter 7
Performance Analysis of SSQ

7.1 Motivation

Simulator Sickness (or cybersickness) is triggered by visual stimuli during exposure in a VE by via HMDs. Different symptoms can be induced during exposure, such as headache, dizziness, eyestrain, blurred vision, etc. (Carnegie and Rhee 2015; Sharples et al. 2008). VR videos provide an immersive experience to users. Besides this, users may experience simulator sickness/cybersickness, which may affect their comfort and viewing experience. To efficiently and accurately self-report the symptoms, an appropriate test method is required. The most popular questionnaire used is the SSQ developed by Kennedy et al. in (1993). This SSQ was designed mainly for measuring simulator sickness experienced by Navy pilots for flight simulator studies. There are differences between simulator and VR systems, and it could be that some of the symptoms relevant for simulator-system studies may not be relevant for VR, and in particular for 360° video. Furthermore, some symptoms belong to more than one category. The SSQ was adapted for VR contents by Kim et al. (2018) and Stone (2017), but not for 360° contents. VR and 360° contents differ in terms of DoF and the way they are captured. Here, VR contains mainly computer-generated imagery and can provide movement to a user in 6 DoF (3 directional and 3 rotational). On the other hand, 360° video is captured from real-life scenes using 360° camera and can provide movement to a user in only 3 DoF (3 rotational). Hence, it could be possible that users may feel less prone to simulator sickness in 360° videos than VR. Therefore, it would be reasonable to investigate if it is efficient to use the SSQ in its current form for 360° videos (Stone 2017; Rebenitsch and Owen 2016). Along similar lines, there arises the question of whether the set of 16 questions may be further reduced when going from a simulator to 360° video, to enable a more dedicated and time-efficient assessment. The resulting simplified version of the SSQ focuses on the symptoms that are relevant for 360° video related studies, with the aim to reduce test time and ultimately enable larger-scale testing.

A. Singla, *Assessment of Visual Quality and Simulator Sickness for Omnidirectional Videos*, T-Labs Series in Telecommunication Services,
https://doi.org/10.1007/978-3-031-49988-3_7

Please note that it is not being argued that the SSQ cannot be used in the case of 360° videos. It may be possible that individual symptoms are less relevant for 360° video.

The data from previously published papers Singla et al. (2017a,b, 2018, 2019) were analyzed in more depth to derive different sets of simplified questionnaires from the SSQ. The main contributions of this chapter are as follows: (1) proposing a simplified SSQ for the self-reporting of symptoms which are relevant for 360° videos, without excluding any relevant symptoms, (2) comparing different versions of simplified SSQs with the state-of-the-art questionnaires, CSQ and VRSQ and with the initial SSQ set of scales, and (3) a simulator sickness database (158 subjects) with subjectively annotated raw scores for different 360° videos and sessions, generated using a wide variety of sequences, different HMDs, different durations of test sessions and with different test designs.

7.2 Experiment

For this study, the data from tests 1–6 is used (see Chap. 3). For all the tests, the Whirligig player was used for the playout of the 360° videos. Table 7.1 lists the information about HMD names, number of participants and outliers, as well as the duration of the VR sessions. The outlier detection was performed on the quality ratings to check the reliability of the test participants, based on ITU-R BT.500-13 (2012) for test 1, and the Pearson correlation coefficient (Tan et al. 2014) for tests 2–6. To find the suitable test method for the assessment of simulator sickness, the data only from reliable subjects were considered. Accordingly, one participant had to be excluded in test 2, one in test 3, four in test 4, five in test 5, three in test 6, and no participant had to be excluded in test 1. Thus, the response from 158 valid

Table 7.1 Information about different tests that were used for the assessment of simulator sickness (Singla et al. 2021)

Test #	HMD name	Number of subjects	Number of outliers	Number of times SSQ filled	Duration of VR session
1 (Singla et al. 2017b)	HTC Vive and Oculus Rift	28	0	24	1 minutes
2 (Singla et al. 2017a)	Oculus Rift	30	1	4	15 minutes
3 (Singla et al. 2018)	Oculus Rift	27	1	4	10 minutes
4 (Singla et al. 2019)	HTC Vive Pro	29	4	4	7 minutes
5 (Singla et al. 2019)	HTC Vive Pro	30	5	4	11 minutes
6 (Singla et al. 2019)	HTC Vive Pro	28	3	4	11 minutes

Table 7.2 Principal component analysis on raw simulator sickness scores for subject ID 1 for test 1 (Singla et al. 2021)

Variance	34.859	19.99	10.613	10.004	8.366
Symptom	PC 1	PC 2	PC 3	PC 4	PC 5
General discomfort	0.588	0.082	−0.124	−0.342	−0.015
Fatigue	0.1	−0.493	−0.313	0.633	−0.086
Headache	0.239	−0.414	0.169	−0.132	0.761
Eye strain	0.236	0.614	−0.437	0.393	0.358
Difficulty focusing	0	0	0	0	0
Salivation increasing	0	0	0	0	0
Sweating	0	0	0	0	0
Nausea	0.455	0.071	−0.097	−0.029	−0.109
Difficulty concentrating	0.256	−0.396	−0.315	0	−0.297
Fullness of head	0.306	0.038	0.679	0.485	0.038
Blurred vision	0.017	0.193	0.243	0.174	−0.268
Dizziness with eye open	0.366	0.014	0.19	−0.073	−0.331
Dizziness with eye closed	0.123	0.017	0.037	−0.141	−0.035
Vertigo	0.123	0.017	0.037	−0.141	−0.035
Stomach awareness	0	0	0	0	0
Burping	0	0	0	0	0

subjects after removing the outliers (see Table 7.1) was considered. All collected data is available online.[1]

For the test 1 and 2, the subjects were asked to fill in the SSQ after watching the stimuli (test 1) or after the test session (test 2). For the test 3, 4, 5, and 6, the SSQ was filled in after the pre-screening, thus before the training session. In addition, for the latter four tests, responses were given after each respective test session, when subjects had viewed a certain number (test dependent) of 360° videos. Note that using the SSQ at the beginning of a test was not done to screen, e.g., for unhealthy participants or to bias their response. It was used with the intention to simply measure their current sickness level, as most of the symptoms can occur to a participant without the use of a simulator or an HMD. For the tests 4, 5, and 6, the SSQ was filled in after the break as well, to investigate the impact of breaks on the scores of simulator sickness. These responses were not considered for the development of the simplified questionnaire.

$$N = (GeneralDiscomfort + IncreasedSalivation + Sweating + Nausea$$
$$+ DifficultyConcentrating + StomachAwareness + Burping) \times 9.54$$
$$O = (GeneralDiscomfort + Fatigue + Headache + EyeStrain$$

[1] https://github.com/Telecommunication-Telemedia-Assessment/360-SimulatorSickness-Data.

$$
\begin{aligned}
&+ DifficultyFocusing \\
&+ DifficultyConcentrating + BlurredVision) \times 7.58 \\
D &= (DifficultyFocusing + Nausea + FullnessOfHead \\
&+ DizzinessWithEyeOpen \\
&+ BlurredVision + DizzinessWithEyeClosed + Vertigo) \times 13.92
\end{aligned}
$$

$$
TS = \left(\frac{N}{9.54} + \frac{O}{7.58} + \frac{D}{13.92}\right) \times 3.74 \tag{7.1}
$$

$$
\begin{aligned}
N &= (GeneralDiscomfort + DifficultyConcentrating) \times 9.54 \\
O &= (GeneralDiscomfort + EyeStrain + DifficultyFocusing \\
&+ DifficultyConcentrating \\
&+ BlurredVision) \times 7.58 \\
D &= (DifficultyFocusing + DizzinessWithEyeOpen \\
&+ BlurredVision) \times 13.92
\end{aligned}
$$

$$
TS = \left(\frac{N}{9.54} + \frac{O}{7.58} + \frac{D}{13.92}\right) \times 3.74 \tag{7.2}
$$

7.3 Simplified Simulator Sickness Questionnaire

7.3.1 Per-subject PCA

PCA is a dimensionality reduction technique that is used to find the essential variables without losing any important information or properties and this reduces the complexity of the data (Kim et al. 2018; Salem and Hussein 2019). PCA has been applied to the SSQ to reduce the number of symptoms (Kim et al. 2018) and to find the factor structure (Bruck and Watters 2011). In addition, factor analysis is done to group the symptoms into different subgroups based on their loading value. In this work, a similar procedure that was followed in the original paper by Kennedy et al. (1993) is applied for factor analysis.

The goal for applying PCA on a per-subject level is to find the most relevant symptoms for each subject and later analyze the number of subjects for whom a particular symptom is relevant. Based on the percentage of occurrence, it was decided whether to include a given symptom in the final questionnaire for each test or not.

In order to keep the PCC of sub-categories/clusters (Nausea (N), Oculomotor (O), and Disorientation (D)) and TotalScore (TS) for the original scoring and the simplified scoring above 95%,[2] the different thresholds for variance explained by the number of Principal Components (PCs) were selected, magnitude for symptom and % of people accordingly.

A PCA was applied to the raw scores of simulator sickness for each subject to get the principal component coefficients and limit the amount of total variance to at least 80% explained by the number of PCs. There is no objective way to find how much variance is needed or how many PCs are sufficient. Therefore, 80% was selected, which is an acceptably large variance. Table 7.2 shows the exemplary principal component coefficients for subject ID_1. It can be seen that five PCs are needed to have a variance of at least 80% for subject ID_1. From Table 7.2 it can be seen that some of the symptoms such as Difficulty Focusing, Salivation Increasing, Sweating, Stomach Awareness, and Burping have almost zero contribution to the PCs. The results further show that different numbers of PCs are needed for each subject, as the variance was fixed, not the number of PCs. So, for some subjects, a higher or lower number of PCs may be needed. Table 7.3 shows the number of PCs required to have a variance of at least 80% and the obtained magnitude for each of the symptoms for each subject. The magnitude of each symptom across the selected PCs using Eq. 7.3 is calculated. Here, J is the number of PCs needed to yield a variance of at least 80%. The obtained magnitude for each of the symptoms will be used for the selection of the most important symptoms.

$$Symptom_i = \sqrt{\sum_{n=1}^{J} PC\_n_i^2} \tag{7.3}$$

The two subjects (ID_11, and ID_27) did not report any of the symptoms, which is why the magnitude of each symptom for these subjects is zero. In a subsequent thresholding, symptoms with a magnitude greater than 0.4 are taken as important for a subject and counted as relevant. Afterwards, it is calculated for how many subjects (in %) a symptom is relevant, which is shown by the last column of Table 7.3. The symptoms, such as Burping, Sweating, and Stomach Awareness, have occurred only for a very few subjects. To keep the simplified TS aligned to the original TS, symptoms being relevant for at least 50% of the subjects are marked as relevant symptoms (highlighted in Table 7.3). Using this criterion, the number of symptoms was reduced from 16 to 6 for test #1, and these six symptoms are General Discomfort, Eye Strain, Difficulty Focusing, Difficulty Concentrating, Blurred Vision, and Dizziness with Eye open. These six symptoms are classified into three sub-categories (N, O, and D), using the same scoring method as described in the original SSQ (Kennedy et al. 1993), see Eq. 7.1. The average N, O, D, and TS over all subjects are calculated for all video sequences based on the Eq. 7.2, for test #1. A correlation analysis is performed based on the simplified scoring and the original

[2] Different thresholds were selected and 95% comes out to be the best.

Table 7.3 Obtained magnitude for each of the symptoms after applying PCA on the raw scores of simulator sickness for each subject for test 1 (Singla et al. 2021)

Subject ID	1	2	3	4	5	6	7	8	9	10	11	12	13	14	15	16	17	18	19	20	21	22	23	24	25	26	27	28	Percentage
No. of PCs	5	5	4	5	4	3	6	3	2	5	N/A	3	4	4	3	3	2	3	1	5	4	3	3	3	3	4	N/A	4	
General discomfort	0.697	0.61	0.529	0.286	0.512	0.766	0.824	0.512	0	0.742	0	0.537	0.531	0.91	0.267	0	0.396	0	0	0.287	0.52	0.489	0.337	0.675	0.305	0.491	0	0.523	57
Fatigue	0.871	0.454	0.981	0.83	0.233	0	0	0.493	0	0.817	0	0.116	0.407	0.861	0	0	0	0.886	0	0.747	0	0.165	0.188	0.31	0	0.666	0	0.468	43
Headache	0.924	0.066	0.321	0.253	0.547	0	0.708	0.399	0	0	0	0	0	0.175	0.182	0	0	0.44	0	0.431	0.287	0	0.207	0.198	0	0.433	0	0.497	25
Eye strain	0.952	0.29	0.743	0.869	0.24	0.1	0.896	0.166	0.61	0.746	0	0	0	0.398	0.835	0.369	0.918	0	0	0.91	0.51	0.437	0.302	0.266	0.545	0.963	0	0.613	50
Difficulty focusing	0	0.518	0.694	0.553	0.75	0.978	0.59	0.237	0.937	0.592	0	0.406	0.559	0	0.657	0.989	0	0	0	0.761	0.477	0.437	0.564	0.653	0.894	0.697	0	0.62	71
Salivation increasing	0	0.91	0.321	0.296	0	0	0.869	0.434	0	0	0	0.565	0	0.453	0	0	0	0	0	0	0.603	0.622	0	0	0	0.333	0	0.289	25
Sweating	0	0	0	0	0	0	0	0.206	0	0	0	0	0	0.154	0	0	0	0	0	0	0.556	0	0	0.213	0	0	0	0.308	4
Nausea	0.484	0.112	0	0	0	0.21	0.22	0.402	0	0	0	0	0.632	0.159	0.505	0	0	0	0	0.694	0.725	0.164	0.624	0.177	0.498	0.577	0	0.555	36
Difficulty concentrating	0.64	0.827	0	0.83	0.6	0.473	0.594	0.314	0	0.738	0	0	0.646	0.68	0.3	0.956	0	0.245	0	0.511	0.441	0.489	0.442	0.574	0.175	0.486	0	0.639	61
Fullness of head	0.89	0.849	0	0.931	0.248	0	0.213	0.348	0	0.62	0	0.537	0	0.551	0	0	0	0	0	0.966	0.371	0.501	0.371	0.394	0	0.135	0	0.587	32
Blurred vision	0.446	0.892	0.908	0.825	0.796	0.981	0.855	0.32	0.866	0.832	0	0.537	0.866	0	0.887	0.977	0.396	0.998	0	0.607	0.746	0.987	0.668	0.949	0.997	0.404	0	0.567	79
Dizziness with eye open	0.534	0.632	0.833	0.505	0	0.465	0.259	0.353	0	0.394	0	0.994	0.702	0.741	0.636	0	0.918	0.695	1	0.508	0.647	0.26	0.489	0.383	0.644	0.613	0	0.474	64
Dizziness with eye closed	0.195	0.686	0	0.66	0	0	0	0	0	0.584	0	0	0.718	0.442	0	0	0	0.695	0	0.508	0.315	0.399	0.489	0	0	0	0	0.552	32
Vertigo	0.195	0.372	0	0.296	0.944	0	0.274	0.325	0	0.87	0	0.806	0.765	0.49	0.478	0.131	0	0	0	0.452	0.677	0.489	0.667	0.442	0.305	0.692	0	0.343	43
Stomach awareness	0	0.126	0	0	0	0	0.944	0.73	0	0	0	0	0.283	0.486	0	0	0	0	0	0	0.31	0.103	0.391	0.326	0.175	0	0	0.53	14
Burping	0	0	0	0	0.904	0	0.836	0.879	0	0	0	0	0	0	0	0	0	0	0	0	0	0	0.391	0	0	0	0	0.111	11

Table 7.4 Variance and magnitude thresholds for different tests (Singla et al. 2021)

Test #	Variance (in %)	Magnitude_Threshold	% of people
1	80	0.4	≥50
2	80	0.2	≥45
3	80	0.3	≥50
4	80	0.25	>40
5	80	0.2	>40
6	80	0.3	>40

scoring. Results show high PCC of 0.969, 0.991, 0.956, and 0.982 for all the three sub-categories of SSQ (N, O, and D) and the Total Score, respectively, for test #1.

Similarly, per-subject PCA was performed on the other five subjective tests and performed all the above-mentioned steps. As described in Sect. 7.2, each test has a different test design and objective; therefore, different thresholds were selected for each test. Table 7.4 lists the different thresholds for magnitude and (%) of people for tests #1–#6. A different simplified questionnaire, referred to as R1–R6 for test #1–test #6, is obtained, respectively, as shown in Table 7.5 and classified the symptoms according to the initial sub-categories of the SSQ (N, O, and D). It can clearly be seen from Table 7.5 that different tests have resulted in different sets of relevant symptoms, except for test #6. The relevant symptoms for tests #5 and #6 are exactly the same. Further, three additional questionnaires (R7, R8, and R9) using questionnaires R1–R6 are derived. The R7 is derived using the union operation on all the questionnaires R1–R6 ($R7 = R1 \cup R2 \cup R3 \cup R4 \cup R5 \cup R6$), resulting in nine symptoms. For R8, the intersection operation is performed and includes only those symptoms that belong to at least five different questionnaires ($R8 = R1 \cap R2 \cap R3 \cap R4 \cap R5 \cap R6, where, Symptom_i \in atleast 5 of \{R1, R2, .. R6\}$), resulting in five symptoms. R9 is derived by performing a union operation between R8 and Dizziness With Eye Open ($R9 = R8 \cup DizzinessWithEyeOpen$), resulting in six symptoms. Dizziness With Eye Open was found to be the relevant symptom only for test #1. This may be explained by the fact that test #1 is quite different from the other five tests in terms of test design and the duration of the videos. Therefore, the symptom Dizziness With Eye Open is included in the R9 questionnaire.

In order to compare the set of questions with the SSQ, the symptoms are categorized in three categories as described in the original paper introducing the SSQ (Kennedy et al. 1993) for all the questionnaires, as shown in Table 7.5. Here, it is assumed that the three clusters from Kennedy et al. (1993) are still valid. For the comparison, the PCC is calculated, so as to identify which of the questionnaires (out of 11) has the maximum correlation with the SSQ. Therein, the N, O, and D analysis only serves as an additional test to determine if the correct symptoms were left out. The same scoring method is used as described in the original SSQ (Kennedy et al. 1993), see Eq. 7.1. From these six tests, nine different questions are derived and two questionnaires (VRSQ and CSQ) are from the state of the art. PCC analysis is conducted between the simplified scoring and the original scoring for all the eleven questionnaires for all the three sub-categories of SSQ (N, O, and D) and the Total Score, respectively, on each of the tests. Table 7.6 shows Pearson Correlation

Table 7.5 Different simplified simulator sickness questionnaires obtained from several tests using PCA (Singla et al. 2021)

Symptoms	R1 (test 1)			R2 (test 2)			R3 (test 3)			R4 (test 4)			R5 (test 5)			R6 (test 6)			R7 (union)			R8 (intersection1)			R9 (intersection2)			VRSQ			CSQ		
	N	O	D	N	O	D	N	O	D	N	O	D	N	O	D	N	O	D	N	O	D	N	O	D	N	O	D	N	O	D	N	O	D
General discomfort	×	×		×	×					×	×		×	×		×	×		×	×		×	×		×	×		×	×				
Fatigue					×			×			×			×			×			×			×			×			×				
Headache											×									×									×			×	
Eye strain		×			×			×			×			×			×			×			×			×			×			×	
Difficulty focusing		×	×		×	×								×	×		×	×		×	×								×	×		×	×
Salivation increasing																																	
Sweating																																	
Nausea																															×		×
Difficulty concentrating	×	×		×	×		×	×					×	×		×	×		×	×		×	×		×	×							
Fullness of head						×			×			×			×			×			×			×			×			×			×
Blurred vision		×	×																	×	×								×	×		×	×
Dizziness with eye open			×																		×						×						×
Dizziness with eye closed																														×			×
Vertigo																														×			×
Stomach awareness																																	
Burping																																	

Table 7.6 Pearson correlation coefficient between N, O, D, and TS of the SSQ and several reduced questionnaires obtained from PCA (Singla et al. 2021)

	Test 1				Test 2				Test 3				Test 4				Test 5				Test 6				Test 7			
Questionnaires	N	O	D	TS	N	O	D	TS	N	O	D	TS	N	O	D	TS	N	O	D	TS	N	O	D	TS	N	O	D	TS
R1	0.969	0.991	0.956	0.982	0.824	0.873	0.646	0.817	0.98	0.998	0.977	1	0.969	0.992	0.991	0.998	0.976	0.991	0.878	0.977	0.973	0.997	0.985	0.999	0.957	0.988	0.985	0.977
R2	0.969	0.991	0.925	0.988	0.824	0.999	0.685	0.977	0.98	0.999	0.941	0.994	0.969	0.99	0.98	0.996	0.976	0.998	0.972	0.994	0.973	0.991	0.985	0.984	0.957	0.997	0.954	0.988
R3	0.875	0.953	0.865	0.941	0.873	0.982	0.203	0.981	0.996	0.997	0.968	1	0.979	0.972	0.757	0.982	0.937	0.987	0.85	0.989	0.897	0.933	0.99	0.95	0.953	0.987	0.967	0.998
R4	0.967	0.943	0.865	0.962	-0.746	0.943	0.203	0.911	0.9	0.995	0.968	0.991	0.959	0.998	0.757	0.991	0.903	0.995	0.85	0.982	0.824	0.947	0.99	0.951	0.948	0.992	0.967	0.995
R5	0.969	0.991	0.925	0.988	0.824	0.999	0.685	0.977	0.98	0.999	0.941	0.994	0.969	0.99	0.98	0.996	0.976	0.998	0.972	0.994	0.973	0.991	0.985	0.984	0.957	0.997	0.954	0.988
R6	0.969	0.991	0.925	0.988	0.824	0.999	0.685	0.977	0.98	0.999	0.941	0.994	0.969	0.99	0.98	0.996	0.976	0.998	0.972	0.994	0.973	0.991	0.985	0.984	0.957	0.997	0.954	0.988
R7 (union)	0.969	1	0.967	0.989	0.824	1	0.938	0.982	0.98	1	0.997	1	0.969	1	0.974	0.992	0.976	1	0.988	0.994	0.973	1	0.989	1	0.957	1	0.985	0.993
R8 (intersection 1)	0.969	0.958	0.865	0.971	0.824	0.985	0.203	0.988	0.98	0.998	0.968	0.994	0.969	0.991	0.757	0.997	0.976	0.996	0.85	0.999	0.973	0.989	0.99	0.99	0.957	0.99	0.967	0.987
R9 (intersection 2)	0.969	0.958	0.92	0.972	0.824	0.985	0.428	0.992	0.98	0.998	0.998	0.999	0.969	0.991	0.912	0.995	0.976	0.996	0.993	0.992	0.973	0.989	0.985	0.989	0.957	0.99	0.99	0.99
VRSQ	0.967	0.993	0.984	0.985	-0.746	0.994	0.963	0.991	0.9	0.999	0.996	0.995	0.959	1	0.982	0.988	0.903	0.997	0.996	0.986	0.824	0.997	0.991	0.99	0.948	1	0.979	0.996
CSQ	*0.95*	*0.945*	*1*	*0.985*	*0.461*	*0.956*	*1*	*0.989*	*0.956*	*0.996*	*1*	*0.999*	*0.307*	*0.997*	*1*	*0.999*	*0.372*	*0.994*	*1*	*0.998*	*0.951*	*0.985*	*1*	*0.959*	*0.901*	*0.991*	*1*	*0.992*

Table 7.7 Pearson correlation coefficient averaged over all tests for different categories of SSQ for several reduced questionnaires obtained using PCA and for VRSQ and CSQ (Singla et al. 2021)

	Correlation value				
Questionnaires	N	O	D	TS	Average_All
R1	0.949	0.974	0.906	0.962	0.947
R2	0.949	0.995	0.915	0.989	0.962
R3	0.926	0.971	0.772	0.974	0.911
R4	0.634	0.97	0.772	0.965	0.835
R5	0.949	0.995	0.915	0.989	0.962
R6	0.949	0.995	0.915	0.989	0.962
R7	**0.949**	**1**	0.975	**0.993**	**0.979**
R8	0.949	0.986	0.772	0.99	0.924
R9	**0.949**	0.986	0.872	0.99	**0.949**
VRSQ	0.634	0.996	0.985	0.989	0.901
CSQ	0.666	0.979	**1**	0.988	0.908

Coefficient for all the tests. Most of the questionnaires have high PCC values for each of the tests except for test #2, where all questionnaires have relatively lower PCC values. R4 and VRSQ have a negative correlation with the Nausea component of the SSQ for the test #2. With the increase in the session duration, the SSQ scores increases. But with R4 and VRSQ, the scores for the Nausea component decrease with time due to the usage of a relatively smaller number of symptoms as compared to the original SSQ.

Table 7.7 shows the average PCC values over all tests and categories of the SSQ (N, O, and D) and TS for each of the questionnaires. It can be clearly seen that R7 (9 symptoms) has the highest PCC with the SSQ as compared to the other questionnaires for N, O, and TS. Other questionnaires, such as R2, R5, and R6 have higher correlations with N, O, D, and TS and contain only 6 symptoms. R9 also has a higher correlation and contains 6 symptoms as well. VRSQ and CSQ have the lowest PCC as compared to the rest of the questionnaires for Nausea category and the average PCC over N, O, D, and TS. The possible reason could be that VRSQ and CSQ are derived using only VR contents, not 360° videos, and both of the questionnaires have used only a very limited VR dataset.

Based on the analysis, the R7 simplified SSQ is recommended, as it has the highest PCC as compared to the state-of-the-art questionnaires VRSQ and CSQ, and the number of symptoms (9) remains the same. Furthermore, R9 simplified SSQ is recommended, as it has a lower number of symptoms, i.e., 6 and still has a higher PCC value. The reason for using R9 instead of R1 or R2 or R5 is that it contains the relevant symptoms, such as Fatigue and Dizziness with Eye Open. Fatigue is relevant when a person is in a VR session for a longer duration of time. Whereas Dizziness with Eye Open is relevant for a subjective test, where contents are being used that contain high amounts of motion.

In order to ascertain whether any frequently occurring or relevant symptom was left out during PCA, the raw scores were averaged over all subjects and

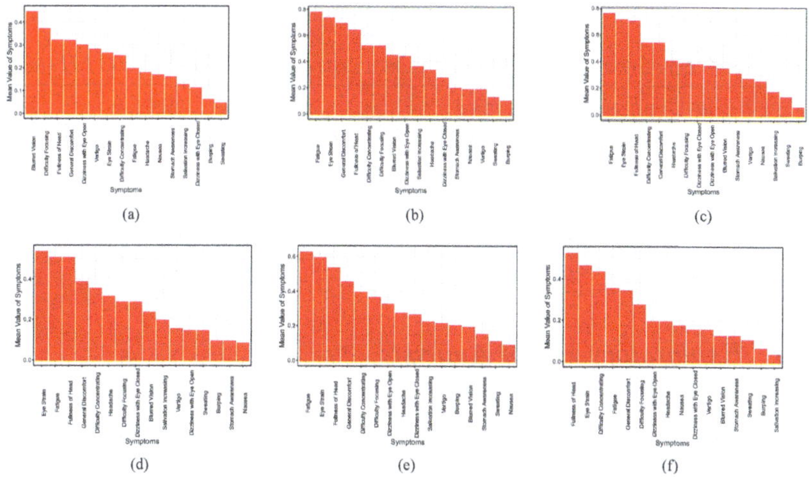

Fig. 7.1 Mean value of all symptoms for several tests (Singla et al. 2021). (**a**) Test 1. (**b**) Test 2. (**c**) Test 3. (**d**) Test 4. (**e**) Test 5. (**f**) Test 6

sessions/videos to obtain the average value of a symptom for each test. Figure 7.1 shows the average value of each symptom for each test. It is interesting to note that the relevant symptoms obtained after applying PCA for each test are contained within the top 8 symptoms having the highest magnitude. Some of the symptoms, such as Nausea, Burping, Sweating, Salivation Increasing, Stomach Awareness, Vertigo, and Dizziness with Eyes closed, were never selected as a relevant symptom and also lie within the bottom 8 symptoms. After identifying the most similar questionnaire w.r.t. SSQ, the next step was to investigate whether these three symptom clusters are still valid for 360° related studies.

$$\begin{aligned}
&Uneasiness\\
&\quad = \frac{(GeneralDiscomfort + Fatigue + EyeStrain + Headache)}{18} \times 100\\
&\qquad + \frac{(DifficultyConcentrating + FullnessOfHead)}{18} \times 100\\
&VisualDiscomfort = \frac{(DifficultyFocusing + BlurredVision)}{6} \times 100\\
&LossofBalance = \frac{(DizzinessWithEyeOpen)}{3} \times 100\\
&TotalScore = \frac{(Uneasiness + VisualDiscomfort + LossOfBalance)}{3}
\end{aligned} \tag{7.4}$$

$$Uneasiness = \frac{(Fatigue + EyeStrain + DifficultyConcentrating + FullnessOfHead)}{12} \times 100$$

$$Lossof Balance = \frac{(GeneralDiscomfort + DizzinessWithEyeOpen)}{6} \times 100$$

$$TotalScore = \frac{(Uneasiness + LossOfBalance)}{2} \quad (7.5)$$

7.3.2 Exploratory Factor Analysis

As explained in Sect. 7.1, due to the differences between VR and simulator usage, it is of relevance to analyze whether the underlying clusters or latent factors initially described by Kennedy et al. (1993) are applicable to simulator sickness in the context of 360° video. To this aim, exploratory factor analysis is conducted to the SSQ scores obtained across the different tests. In order to apply factor analysis, the raw simulator sickness scores are averaged across each subject for each symptom and PVS/session. These averaged out values for each test and the corresponding session/PVS are stacked in a single file.

The Kaiser-Meyer-Olkin (KMO) test was performed to determine if the data used for factor analysis is suitable or not. In this study, the KMO value found is 0.82, which means that the data can be considered suitable for factor analysis (Kaiser 1970; Kaiser et al. 1974).

Exploratory Factor Analysis (EFA) was first performed on the R7 (Union) and R9 (Intersection 2) questionnaires, including varimax rotation. To find out how many factors are sufficient, the hypothesis test is performed, suggesting three and two factors are sufficient for the case of R7 and R9, respectively. The variance explained by the three factors (R7) and two components (R9) is 85.4% and 86.1%, respectively, of the total variance. Tables 7.8 and 7.9 show the factor loadings for the varimax orthogonal rotation for individual symptoms for R7 and R9 questionnaires. For the R7 questionnaire, factors with less than 0.53 factor loadings were removed, resulting in a simple structure, where each symptom belongs to one factor only, which eases the naming of the factors. Factor 1 loads mainly on aspects of uneasiness such as General Discomfort, Fatigue, Eye Strain, Headache, Difficulty Concentrating and Fullness of head. Factor 2 loads mainly on visual-discomfort-related symptoms such as Difficulty Focusing and Blurred Vision. Factor 2 could also be referred to as visual quality factors, related with both the quality-rating task present in all six tests as well as the types of degradations included in many of these (different resolutions, encoding bitrates, HMDs). Factor 3 loads only on Loss of Balance and Dizziness with eyes open. One could summarize the factor labels for R7 as Factor 1: Uneasiness, Factor 2: Visual Discomfort and Factor 3: Loss of Balance. A simple average method was used based on Kim et al. (2018) to calculate

Table 7.8 Varimax factors for reduced SSQ (Union) (Singla et al. 2021)

Symptom	Factor 1	Factor 2	Factor 3
General discomfort	**0.701**	0.385	0.52
Fatigue	**0.953**	0	0.148
Headache	**0.826**	0.116	0.298
Eye strain	**0.938**	0.136	0
Difficulty focusing	0.341	**0.923**	0.166
Difficulty concentrating	**0.786**	0.222	0.343
Fullness of head	**0.897**	0	0.224
Blurred vision	−0.101	**0.801**	0.202
Dizziness with eye open	0.298	0.404	**0.766**

Table 7.9 Varimax factors for reduced SSQ (intersection 2) (Singla et al. 2021)

Symptom	Factor 1	Factor 2
General discomfort	0.595	**0.747**
Fatigue	**0.916**	0.269
Eye strain	**0.913**	0.223
Difficulty concentrating	**0.720**	0.518
Fullness of head	**0.867**	0.317
Dizziness with eye open	0.176	**0.887**

Uneasiness, Visual Discomfort, Loss of Balance, and total score as shown in Eq. 7.4 for the R7 questionnaire.

For the R9 questionnaire, factors with less than 0.60 factor loadings were removed from each component, resulting in a simple structure, where each symptom belongs to one factor only, easing the naming of the factors. Factor 1 loads mainly on aspects of uneasiness such as Fatigue, Eye Strain, Difficulty Concentrating and Fullness of head. Factor 2 loads only on loss of balance, General Discomfort and Dizziness with eyes open. One could summarize the factor labels for R9 as Factor 1: Uneasiness and Factor 2: Loss of Balance. Uneasiness, Loss of Balance and the total score are calculated based on a simple average method described in Kim et al. (2018) as shown in Eq. 7.5 for the R9 questionnaire.

Cronbach's Alpha (α) is measured, which is a measure of the internal consistency. For the R7, Cronbach's Alpha is 0.91 and for R9 it was found to be 0.93, which signifies that the questionnaires examined here show sufficient reliability.

Thus, the outcomes from this study do not negate the factor structure from the study conducted by Kennedy et al. (1993), especially for the R7 questionnaire. Both of the studies point to the same factors, namely loss of balance (Disorientation), visual discomfort (Oculomotor) and uneasiness (Nausea), contributing to the assessment of simulator sickness, but contains different symptom clusters.

Further, EFA was performed on the original SSQ using the previously published data in order to check if the same set of symptom clusters can be obtained for this test as those described for the original SSQ (Kennedy et al. 1993). Table 7.10 shows the factor loadings for the varimax orthogonal rotation for individual symptoms for the SSQ questionnaires. The variance explained by the three factors is 73.5% of

Table 7.10 Varimax factors for SSQ (original) (Singla et al. 2021)

Symptom	Factor 1 (N)	Factor 2 (D)	Factor 3 (O)
General discomfort	**0.68**	**0.583**	**0.303**
Fatigue	**0.968**	0.112	0
Headache	**0.819**	**0.349**	0
Eye strain	**0.918**	0	0.205
Difficulty focusing	0.268	**0.403**	**0.872**
Salivation increasing	**0.619**	**0.317**	**0.317**
Sweating	**0.518**	0.151	0
Nausea	0.2	**0.755**	0.28
Difficulty concentrating	**0.759**	**0.426**	0.178
Fullness of head	**0.895**	0.22	0
Blurred vision	−0.138	**0.347**	**0.728**
Dizziness with eye open	0.299	**0.802**	0.228
Dizziness with eye closed	**0.8**	**0.361**	0
Vertigo	0	**0.93**	0.25
Stomach awareness	**0.442**	**0.67**	0.184
Burping	**0.444**	0	0

the total variance. To find which of the symptom belongs to which cluster, factors with less than 0.30 factor loadings were removed from each factor. Here, the same threshold as defined in the SSQ (Kennedy et al. 1993) is kept. It can be seen that the resulting symptom clusters are different from the ones defined by Kennedy et al. for N, O, and D. For this case of 360° video, a different set of symptoms was found, belonging to different factors.

7.3.3 Discussion

The subjective tests (#1–#6) that were considered for developing the reduced-set questionnaire were originally targeting research on video quality. From Fig. 7.1, it can clearly be seen that the strongest effect on participants is that of Blurred vision, Difficulty Focusing, Eye Strain, Difficulty Concentrating and General Discomfort. Due to the variation in resolution and bitrate that have been chosen for test stimulus generation, the SSQ itself may address certain aspects of simulator sickness which are not necessarily simulator sickness issues, but may be due to the quality-related characteristics investigated in those studies. Then, these quality-related issues may translate into simulator sickness issues when rating corresponding symptoms in the simulator sickness questionnaire. To further analyze this hypothesis, the data from a further test (Test #7) is used, which is publicly available. That test focused only on collecting viewing behavior from subjects when watching 360° videos on an HMD, collecting simulator sickness data for complementary analysis. In that test, no variation in resolution and bitrate was applied (Fremerey et al. 2018). Inspection of this additional test revealed that users still reported high scores for Blurred vision, Difficulty Focusing, Eye Strain, Difficulty Concentrating and General Discomfort. Therefore, it can be assumed that the quality-related issues introduced in the tests #1–#6 may not have been translated into simulator sickness issues when rating corresponding symptoms in the simulator sickness questionnaire. Furthermore, all the reduced questionnaires (R1–R9), VRSQ and CSQ are compared using PCC analysis including Test #7. It can be seen from Table 7.6 that all the questionnaires including VRSQ and CSQ show high correlations with N, O, D, and TS. This observation supports the conclusion that the SSQ or a corresponding simplified questionnaire can be instantiated in the same way both for 360° video tests, including dedicated media-quality-related questions, and for tests without focus on visual quality. Also, the proposed R7 questionnaire can be used in the contexts that use 360° videos and 360° audiovisual stimuli, as the R7 questionnaire is designed using all tests (#1–#6) and contains all of the symptoms from test 1, which used audiovisual stimuli. Hence, for the range of conditions that have been applied, it is considered being useful.

Please be aware that there are some limitations of the simplified questionnaire. It is noted that the results in this chapter are obtained by simulating the usage of the reduced-set questionnaires, simply omitting the questions that are proposed to be left out for 360° video testing. The proposed 9-attribute questionnaire will need to be verified in concrete tests with the reduced set of questions in the future.[3] Also, the simplified questionnaire was developed using a specific set of 360° videos, and the duration of an individual test session was short.

7.4 Conclusions

In this study, an attempt is made to develop and validate efficient questionnaires for the assessment of simulator sickness in the context of 360° videos. To this aim, the data of six previously published studies conducted by our group were analyzed. This consists of results obtained from more than 150 participants covering 360° videos with wide ranging spatio-temporal complexity tested with different HMDs with different test designs and paradigms. From this, nine different reduced questionnaires were derived from the original SSQ by applying PCA (per-subject) on the results obtained from the six different subjective tests. A PCC analysis was conducted to check the applicability of the proposed reduced questionnaires in comparison to the SSQ and how they compare with the CSQ Stone (2017) and VRSQ Kim et al. (2018) proposed in literature. Experimental results suggest that the proposed R7 questionnaire shows the highest PCC with the SSQ in terms of the

[3] This is due to the current pandemic and associated lockdown of our institution, which has limited the ability to run dedicated additional laboratory tests.

initial clusters. Further, results of factor analysis showed that nine symptom-related attributes appear to be sufficient to represent the three dimensions of SSQ, as some symptoms included in the SSQ are not relevant for 360° video studies and can be expected to capture all the relevant symptoms, just as the SSQ does, and hence can be used as an alternative to the SSQ for studies related to 360° videos. Therefore, based on the analysis, it is proposed to use the simplified SSQ as it contains a lower number of questions (9 instead of 16) with higher correlation with the original SSQ as compared to VRSQ and CSQ. To conclude, the results from this study imply that the simplified version of the SSQ focuses on the symptoms that are relevant for 360° video related studies and can be used efficiently, enabling larger-scale testing.

For future efforts, it is proposed to further evaluate the SSQ with a broader range of contents such as VR and CGI-based 360° videos, and also to construct an even more holistic yet sufficient subjective measurement tool. Furthermore, the duration of the VR session is extended to enhance the validation of the different questionnaires.

References

S. Bruck, P.A. Watters, The factor structure of cybersickness. Displays **32**(4), 153–158 (2011). Visual Image Safety. ISSN: 0141-9382. https://doi.org/10.1016/j.displa.2011.07.002. http://www.sciencedirect.com/science/article/pii/S014193821100059X

K. Carnegie, T. Rhee, Reducing visual discomfort with HMDs using dynamic depth of field. IEEE Comput. Graph. Appl. **35**(5), 34–41 (2015). ISSN: 0272-1716. https://doi.org/10.1109/MCG.2015.98

S. Fremerey et al., AVtrack360: An open dataset and software recording people's head rotations watching 360-degree; videos on an HMD, in *Proceedings of the 9th ACM MMSys*. Amsterdam, Netherlands (2018). ISBN: 978-1-4503-5192-8. https://doi.org/10.1145/3204949.3208134. http://doi.acm.org/10.1145/3204949.3208134

ITU-R BT.500-13, *Recommendation ITU-R BT.500-13: Methodology for the Subjective Assessment of the Quality of Television Pictures* (2012)

H.F. Kaiser, A second-generation little Jiffy. Psychometrika **35**(4), 401–15 (1970)

H.F. Kaiser, J. Little, I.V. Mark, Educ. Psychol. Meas. **34**, 11–17 (1974)

R.S. Kennedy et al., Simulator sickness questionnaire: an enhanced method for quantifying simulator sickness. Int. J. Aviat. Psychol. **3**(3), 203–220 (1993)

H.K. Kim et al., Virtual reality sickness questionnaire (VRSQ): motion sickness measurement index in a virtual reality environment. Appl. Ergon. **69**, 66–73 (2018). ISSN: 0003-6870. https://doi.org/10.1016/j.apergo.2017.12.016. http://www.sciencedirect.com/science/article/pii/S000368701730282X

L. Rebenitsch, C. Owen, Review on cybersickness in applications and visual displays. Virtual Real. **20**(2), 101–125 (2016). ISSN: 1359-4338. https://doi.org/10.1007/s10055-016-0285-9

N. Salem, S. Hussein, Data dimensional reduction and principal components analysis. Proced. Comput. Sci. **163**, 292–299 (2019). 16th Learning and Technology Conference 2019 Artificial Intelligence and Machine Learning: Embedding the Intelligence. ISSN: 1877-0509. https://doi.org/10.1016/j.procs.2019.12.111. http://www.sciencedirect.com/science/article/pii/S1877050919321507

S. Sharples et al., Virtual reality induced symptoms and effects (VRISE): comparison of head mounted display (HMD), desktop and projection display systems. Displays **29**(2), 58–69 (2008). Health and Safety Aspects of Visual Displays. ISSN: 0141-9382.

https://doi.org/10.1016/j.displa.2007.09.005. http://www.sciencedirect.com/science/article/pii/S014193820700100X

A. Singla et al., Comparison of subjective quality evaluation for HEVC encoded omnidirectional videos at different bit-rates for UHD and FHD resolution, in *Proceedings of the on Thematic Workshops of ACM Multimedia*. Mountain View, California, USA (2017a). ISBN: 978-1-4503-5416-5. https://doi.org/10.1145/3126686.3126768. http://doi.acm.org/10.1145/3126686.3126768

A. Singla et al., Measuring and comparing QoE and simulator sickness of omnidirectional videos in different head mounted displays, in *9th International Conference on QoMEX* (2017b), pp. 1–6. https://doi.org/10.1109/QoMEX.2017.7965658

A. Singla, W. Robitza, A. Raake, Comparison of subjective quality evaluation methods for omnidirectional videos with DSIS and Modified ACR. Electron. Imag. **2018**, 1–6 (2018). https://doi.org/10.2352/ISSN.2470-1173.2018.14.HVEI-525

A. Singla, W. Robitza, A. Raake, Comparison of subjective quality test methods for omnidirectional video quality evaluation, in *2019 IEEE 21st International Workshop on Multimedia Signal Processing (MMSP)* (2019), pp. 1–6

A. Singla et al., Assessment of the simulator sickness questionnaire for omnidirectional videos, in *2021 IEEE Virtual Reality and 3D User Interfaces (VR)* (2021), pp. 198–206

W.B. Stone III, *Psychometric Evaluation of the Simulator Sickness Questionnaire as a Measure of Cybersickness*. Doctoral dissertation, Iowa State University, 2017

T.K. Tan et al., HEVC subjective video quality test results, in *IBC2014 Conference, 2014*. Amsterdam, Netherlands Conference , 11–15 Sept, 2014. United Kingdom: Institution of Engineering and Technology (2014). ISBN: 978-1-84919-927-8. https://doi.org/10.1049/ib.2014.0029

Chapter 8
Evaluation of Different Constructs: Media Quality, Simulator Sickness, and Presence

8.1 Experimental Setup

In this subjective test (Test 12), videos from test 1 were used. The test consisted of two sessions, and in each, a different test protocol was applied. In each session, after each video, subjects first were asked to rate quality (Q). The sessions differed in either simulator sickness or presence that was rated using the single scale, "short" questionnaire (referred to as SSQ_S and PQ_S in the following), while the respective other construct was rated using the well-established "long" version of the questionnaire (SSQ_L, PQ_L). To control for session-order effects, the subjects were divided equally into two groups evaluating the same video sequences. One group was to provide ratings in the order [Q, PQ_S, SSQ_L] in the 1st and [Q, SSQ_S, PQ_L] in the 2nd session, with the two sessions exchanged for the other group. Based on the approach, the results can be analyzed either in terms of a between-subject (for the same session) or within-subject design (comparing sessions one and two) or in a mixed way.

8.1.1 Technical Setup and Equipment

The HTC Vive Pro was used as the display device and the *Whirligig* player as the playout interface. The resolution and FOV of the used HMD are 2880×1600 and 110° horizontally, respectively. The HMD was connected to a desktop PC equipped with an SSD, NVIDIA GTX1080 graphics card, and a current CPU (Intel i7, 6700).

A. Singla, *Assessment of Visual Quality and Simulator Sickness for Omnidirectional Videos*, T-Labs Series in Telecommunication Services,
https://doi.org/10.1007/978-3-031-49988-3_8

8.1.2 Test Method

The 5-point ACR scale (ITU-T Rec. P.910 2008) was used for evaluating the integral quality of the 360° videos. The duration of the PVSs was 60 s so that the subjects had sufficient time to experience simulator sickness and presence. Subjects were instructed not to consider possible stitching and ghosting artifacts while rating the integral quality of the videos.

In this experiment, 30 subjects (16 females and 14 males) participated, with an average age of 26.5 years and a median age of 26 years. The range of subjects' age varied from 21 to 33 years. Subjects were divided equally into two groups as described in Sect. 8.1. Prior to the experiment, each test participant was screened for correct visual acuity using Snellen charts (20/25) (ITU-T P.915 2016) and for color vision using Ishihara charts (Ishihara 1997). Subsequently, written instructions were provided to the test subjects explaining the test-related tasks. A training session was conducted in order to familiarize the subjects with the testing procedure and adjusting the HMD according to their head size.

Only one subject performed the test at a time. The subjects were asked to sit on a revolving chair so as to enable them to watch the 360° videos by rotating in any direction. Each subject had to rate 24 PVSs (6 video sequences over 2 sessions in 2 resolutions, full-factorial). First, subjects were asked to view a video of 60 s, and depending on the questionnaires (either simulator sickness or presence), they were given 1–2 minutes to rate the video quality and to fill in the additional questionnaires (SSQ or PQ). Then, a 60 s long pause was given to reduce the effect of one video over the other, especially on presence and simulator sickness evaluation. AVRateNG[1] was used and adapted for recording the user's responses, running on an external tablet device. After viewing an individual video sequence on the HMD, the test subjects took off the HMD and attended to the tablet for giving their ratings.

8.2 Results

8.2.1 Audiovisual Quality

To check the reliability of subjects, outlier detection was performed based on ITU-R BT.500-13 (2012). In this test, no outlier was found, leading to 30 individual scores per test stimulus. MOS and corresponding 95% CIs were computed for each of the test conditions (resolution, session, and content) according to ITU-R Rec. BT.500-13.

[1] https://github.com/Telecommunication-Telemedia-Assessment/avrateNG.

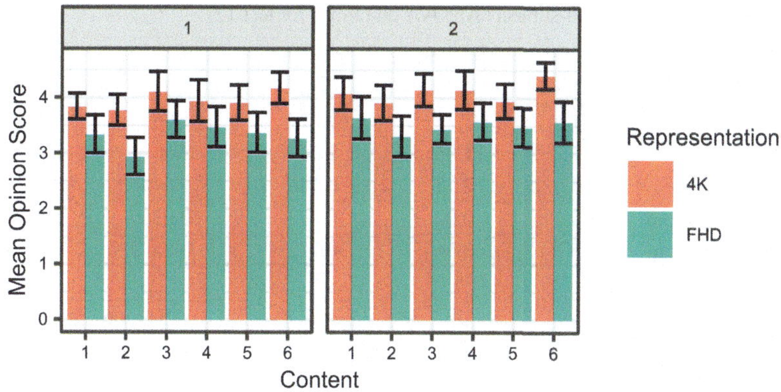

Fig. 8.1 MOS with corresponding CIs for different sessions for 4K and FHD representations for test 12 (Singla et al. 2019)

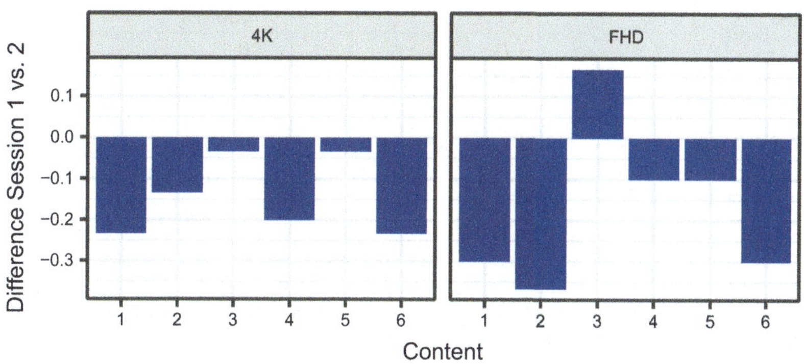

Fig. 8.2 Difference in MOS between sessions for 4K and FHD representation, in terms of MOS (session 1) – MOS (session 2)

Figure 8.1 shows the MOS for all the video sequences for the two different sessions 1 and 2, separately per resolution. It should be noted that these videos were directly downloaded from YouTube and, hence, the bitrate associated with each resolution for different videos was different owing to the difference in content characteristics. As expected, results show that 4K provides better integral quality as compared to FHD resolution and the associated, an effect that was already observed in test 1 of Chap. 5 (Fig. 5.1). As already pointed out in Chap. 5 (Sect. 5.1.3) that not all differences between contents were statistically significant, as can also be seen from the plot in Fig. 8.1. Obviously, some source videos were better suited to convey resolution and bitrate differences than others. One of the goals of this study was to understand the rating behavior of subjects for the same videos at different points in time. Figure 8.2 depicts the difference between MOS values for session # 1 and #2 for 4K and FHD. Negative bars mean that MOS is higher in session #2 than #1. It can be seen that subjects rated the integral quality of video sequences higher

Table 8.1 Results from within-subjects ANOVA on ratings for test 12

	DF	Sum Sq	Mean Sq	*F*	*p*
Session	1	4.36	4.35	5.25	0.022
Resolution	1	67.22	67.22	81.06	<0.01
Content	5	11.05	2.21	2.66	0.02
Session × content	5	2.73	0.546	0.04	0.65
Session × resolution	1	0.02	0.022	0.026	0.87
Resolution × content	5	3.56	0.712	0.8589	0.50

Table 8.2 Comparison of average MOS between session #1 and #2 for test 12

Session	4K	FHD
#1	3.95	3.32
#2	4.09	3.49

Table 8.3 Pearson correlation coefficient and RMSE between different groups for session #1 and #2 for test 12

Session	PCC	RMSE
#1	0.73	0.33
#2	0.84	0.20

Table 8.4 Comparison of average MOS between HTC Vive and Oculus Rift (Singla et al. 2017)

Device	4K	FHD
HTC Vive	3.75	3.16
Oculus Rift	3.62	3.02

in the second session as compared to the first session, irrespective of the resolution, except for video sequence #3 at FHD resolution, where integral quality is higher in session #1.

A within-subjects ANOVA was carried out to statistically evaluate the impact of the experimental factors (resolution associated with bitrates, session, and video sequence) on individual ratings. Results (see Table 8.1) indicate that resolution has a significant contribution to the individual ratings with p-values < 0.01. The session and video sequence have a slight effect ($p = 0.02$).

Table 8.2 shows the average MOS over all video sequences for 4K and FHD resolutions. It is evident that session #2 has higher MOS values than session #1 irrespective of the resolution. It is interesting to note that the difference between the average MOS value of session #1 and #2 for 4K and FHD is similar to 0.63 and 0.60, respectively. Furthermore, Table 8.3 shows the Pearson correlation coefficient and Root Mean Square Error (RMSE) between different groups for several sessions. It is clear from the results that for session #2, the value of correlation coefficient (0.84) is higher than session #1 (0.73). Also, the RMSE value is lower for session #2 than for #1. The average MOS values for HTC Vive Pro (see Table 8.2, test 12) were compared with HTC Vive and Oculus Rift (test 1, see Table 8.4) (Singla et al. 2017). Table 8.4 shows the average MOS value for HTC Vive and Oculus Rift. It is clear from the results that HTC Vive Pro provides better integral quality as compared to the other two HMDs. It is expected because of the higher display resolution.

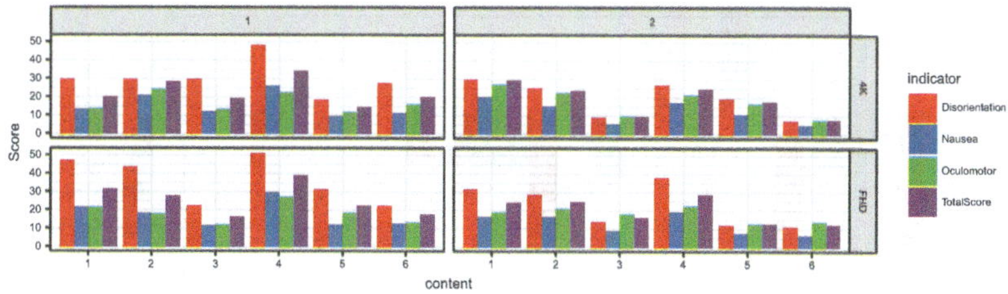

Fig. 8.3 Simulator sickness scores obtained using SSQ_L for different groups and resolutions for test 12

8.2.2 *Evaluation of Simulator Sickness*

While evaluating the integral quality of 360° videos in an HMD, users may experience symptoms of simulator sickness such as nausea, dizziness, vertigo, sweating, etc. Kennedy et al. (1993), Hale and Stanney (2014) and Balk et al. (2013). Hence, evaluating simulator sickness is essential as it forms an important part of the user's experience. For evaluating simulator sickness, the "Long" Simulator Sickness Questionnaire (SSQ_L) proposed in Kennedy et al. (1993) and the "Short" Simulator Sickness Questionnaire (SSQ_S) proposed in Tran et al. (2017) was used.

8.2.2.1 Long Simulator Sickness Questionnaire

Figure 8.3 shows the average simulator sickness scores obtained using SSQ_L for all video sequences and for different groups and resolutions. Results suggest that video #3, #5, and #6 have generally lower simulator sickness scores irrespective of the resolution and group. In video sequence #3, #5, and #6, the motion is lower as compared to video sequence #1, #2, and #4. These videos (#1, #2, and #4) also have a higher value of TI (see Fig. 3.3). Hence, one possible reason for lower simulator sickness in these videos could be attributed to the motion related to camera movement and motion in the content itself. The same effect was observed in Sect. 6.1 (Fig. 6.1). Also, it can be seen from the results that disorientation scores, in general, have the highest scores.

In order to study the impact of resolution[2] on the scores of simulator sickness, the difference between the sickness scores of 4K and FHD is obtained. Figure 8.4 shows the difference between sickness scores for 4K and FHD resolution averaged over the groups. Negative bars mean that users are less prone to simulator sickness in 4K resolution than FHD. It is evident from the figure that users felt less prone

[2] The videos used in this study were directly downloaded from YouTube. Consequently, the bitrate associated with each resolution varied across different videos, reflecting the differences in content characteristics.

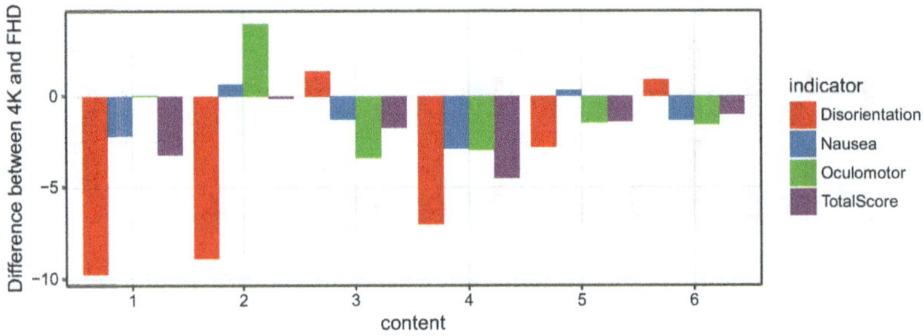

Fig. 8.4 Difference in simulator sickness scores obtained using SSQ_L averaged over groups for different video sequences for test 12

Table 8.5 Results from ANOVA on TotalScore for test 12

	DF	Sum Sq	Mean Sq	*F*	*p*
Resolution	1	23.73	23.734	0.864	0.395
Content	5	1023.60	204.72	7.45	0.022
Resolution × content	5	12.44	2.488	0.09	0.99

to simulator sickness scores while watching the 360° videos in 4K resolution. The same results were achieved in Sect. 6.1 (Fig. 6.2).

Finally, a statistical evaluation was carried out to find the impact of experimental factors such as resolution associated with bitrates, and video sequence on user ratings. For this, an ANOVA was performed on the TotalScores with resolution and video sequence as independent factors. Results (see Table 8.5) indicate that video sequence had a slight effect to the SSQ TotalScore with p-value $= 0.02$. Here, resolution does not seem to have a significant impact on the TotalScore, which is different from the results obtained in Sect. 6.1. However, the impact of resolution on TotalScore is still observable.

8.2.2.2 Short Simulator Sickness Questionnaire

Similar to this test, Tran et al. (2017) also uses a short questionnaire to evaluate simulator sickness among subjects. Figure 8.5 shows the average Short SSQ scores for all the video sequences, sessions, and resolutions. It is clear from the results that videos #1, #2, and #4 lead to higher simulator sickness scores irrespective of the resolution and group. The same results are also obtained by using the Long SSQ. Users are less prone to simulator sickness in FHD than 4K irrespective of the group and video, except video sequence # 2 and #3 which is opposite from the results obtained using the Long SSQ and in Sect. 6.1.

Further, an ANOVA was performed on the individual ratings to find the impact of experimental factors such as resolutions and video sequences on user ratings.

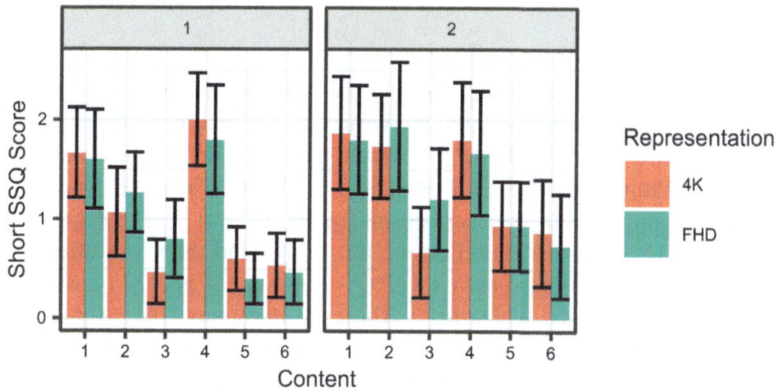

Fig. 8.5 Mean scores obtained using SSQ_S for different sessions and representations. Group #1 filled in SSQ_S in session 2 and group #2 filled in SSQ_S in session 1

Table 8.6 Results from ANOVA on short SSQ ratings for test 12

	DF	Sum Sq	Mean Sq	F	p
Resolution	1	0.100	0.10	0.1143	0.735
Content	5	87.86	17.57	20.09	< 0.01
Resolution × content	5	4.1	0.82	0.93	0.45

Table 8.7 Pearson correlation coefficient between SSQ_S and individual sub-constructs of SSQ_L averaged across all subjects

Group	Short SSQ and Nausea	Short SSQ and Oculomotor	Short SSQ and disorientation	Short SSQ and totalScore
#1	0.70	0.59	0.68	0.69
#2	0.92	0.81	0.89	0.89

Results (see Table 8.6) show that video sequence has a significant contribution to the SSQ score with p-value < 0.01.

One of the goals was to study whether the single-scale SSQ_S is sufficient as compared to the Long SSQ_L for evaluating simulator sickness. The Pearson Correlation Coefficient (PCC) was calculated between SSQ_S and individual sub-constructs of SSQ_L, see Table 8.7. SSQ_S is highly correlated with Nausea scores (Group #1: 0.70 and Group #2: 0.92) and worst correlated with Oculomotor scores (Group #1: 0.59 and Group #2: 0.81). Furthermore, it can be seen that SSQ_S is moderately correlated with the TotalScore (Group #1: 0.69 and Group #2: 0.89). In general, the correlation for the second group is higher than for the first group. The possible reason could be as Group #2 had filled in the SSQ_S in session #1 and SSQ_L in session #2. Therefore, the subjects in this group were trying to correlate most/all of the symptoms of simulator sickness with only one question (SSQ_S) which was not the case with Group #1.

8.2.3 *Presence*

Omnidirectional videos allow users to interact with the environment, which gives them a perception of being present in the virtual world/environment. For measuring the presence in a VE, the "Long" Presence Questionnaire (PQ_L) was used as proposed in Witmer and Singer (1998) and a "Short" Presence Questionnaire (PQ_S) with a single scale as described in Tran et al. (2017). The PQ_L (version 2) consists of 32 questions. Based on the test design and the target effect (e.g., a sense of involvement in the scene), the PQ_L was adapted and 22 questions were selected that fit the goals of this test.[3] These 22 questions[4] (see Appendix B.4) are further classified into 4 different categories: control, sensory, distraction, and realism factors. Subjects had to rate each factor on a 7-point scale. For thePQ_S, subjects had to rate on a single 5-point scale (1: absolutely no sense of presence and 5: a true sense of presence as in a real environment).

8.2.3.1 Long Presence Questionnaire

Figure 8.6 shows the average presence scores for all video sequences computed for both 4K and FHD in both the sessions. For a sense of presence in a VE, the values for the control, realism, and sensory factors should be high and the value of the distraction factor should be low. Results indicate that video #2 provided the least sense of presence as the scores for control, realism, and sensory factors are the lowest compared to other video sequences irrespective of the resolutions[5] and

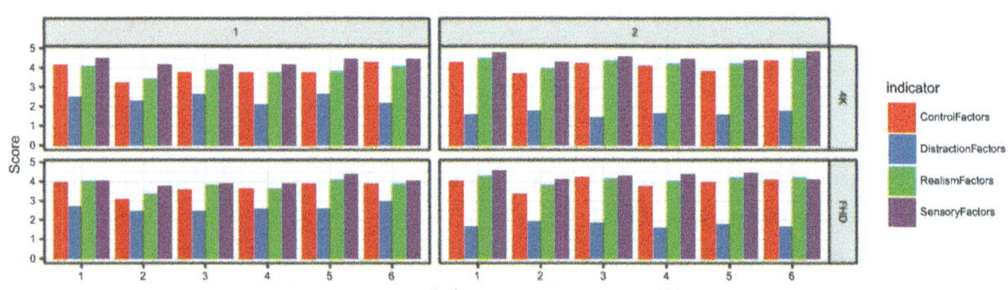

Fig. 8.6 Presence scores aggregated over the four sub-dimensions for different sessions and resolutions for test 12. Group #1 filled in PQ_L in session 2 and group #2 filled in PQ_L in session 1

[3] For example, how well could you examine objects from multiple viewpoints? This question was removed as in this test, subjects could examine objects from only a single viewpoint.

[4] Please note that this was done to make the tests manageable and no formal validation of the reduced set of scales was performed.

[5] The videos used in this study were directly downloaded from YouTube. Consequently, the bitrate associated with each resolution varied across different videos, reflecting the differences in content characteristics.

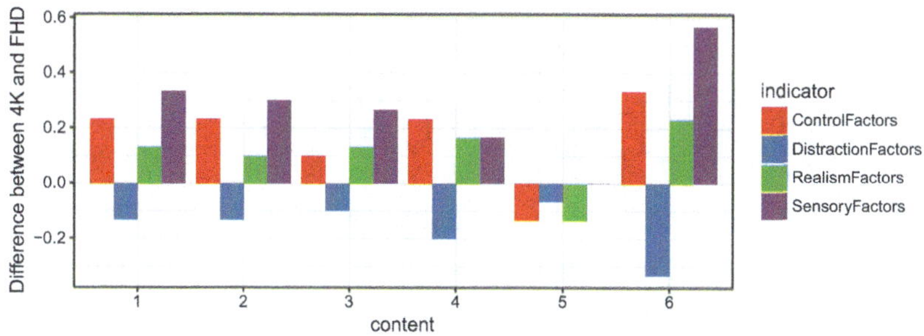

Fig. 8.7 Difference in presence scores obtained using PQ_L averaged over groups for different video sequences for test 12

Table 8.8 Results from ANOVA on realism score for test 12 (Singla et al. 2019)

	DF	Sum Sq	Mean Sq	*F*	*p*
Resolution	1	0.06	0.06	44.02	0.001
Content	5	0.91	0.182	120.21	0.022
Resolution × content	5	0.07	0.015	10.3	0.011

sessions. One possible reason could be that, in video #2, there is much higher motion because of the camera movement and motion in the content itself, which may result in an increased simulator sickness, thereby reducing the chances for a better feeling of presence. This argument can be corroborated from the simulator sickness values for video #2 shown in Fig. 8.3; videos #1 and #6 have the highest scores for control, realism, and sensory factors irrespective of the sessions at 4K resolution. In order to investigate the impact of resolution on the presence scores, the difference between the presence scores of 4K and FHD is obtained. Figure 8.7 shows the difference between presence scores averaged over groups for 4K and FHD resolution. Positive bars for the control, realism, and sensory factors mean that users perceived a higher degree of presence in 4K resolution than FHD. Negative bars for distraction factor mean that the users felt less distracted in 4K than FHD. It is clear from the results that users perceived a higher degree of presence while watching omnidirectional videos in 4K than FHD resolution for all the video sequences except video sequence #5.

To statistically evaluate the impact of experimental factors (resolutions and video sequences) on realism score, an ANOVA was carried out. Results (see Table 8.8) suggest that resolution has a significant impact on the scores of realism with p-values < 0.01 and there is a slight interaction between resolution and content with p-value 0.01.

8.2.3.2 Short Presence Questionnaire

Furthermore, the Short PQ_S is also used for evaluating the presence among test participants in Tran et al. (2017). Figure 8.8 shows the average presence scores

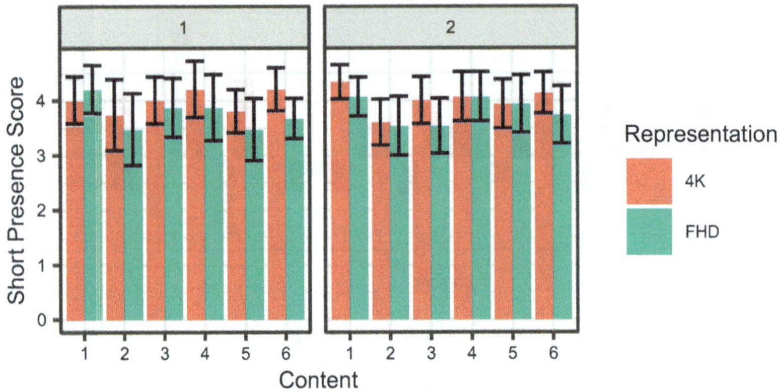

Fig. 8.8 Mean scores obtained using PQ_S for different sessions and representations. Group #1 filled in PQ_S in session 1 and group #2 filled in PQ_S in session 2

Table 8.9 Results from ANOVA on short PQ ratings for test 12

	DF	Sum Sq	Mean Sq	F	p
Resolution	1	4.22	4.22	4.75	0.029
Content	5	12.12	2.42	2.73	0.019
Resolution × content	5	1.65	0.33	0.37	0.86

Table 8.10 Pearson correlation coefficient between short PQ_S and long PQ_L averaged over PVSs

Group	Short PQ_S and control factors	Short PQ_S and distraction factors	Short PQ_S and realism factors	Short PQ_S and sensory factors
#1	0.79	−0.07	0.65	0.62
#2	0.64	−0.43	0.63	0.70

(PQ_S) for all the video sequences and for different resolutions and sessions. Results show that video sequence # 2 has provided the least sense of presence irrespective of the test session at 4K resolution. Whereas, video sequences # 6 and #1 have provided the maximum sense of presence in session #1 and #2, respectively. Similar results are also obtained when using long PQ (PQ_L), as described in Sect. 8.2.3.1. The effect of resolution can also be seen from the short presence scores. 4K resolution generally has higher presence scores than FHD.

To statistically evaluate the impact of resolution and content, an ANOVA was carried out. Results (see Table 8.9) indicate that resolution and content have a slight contribution to the short PQ_S scores with p-values < 0.029 and < 0.019, respectively.

In order to find out to what degrees PQ_S and PQ_L are correlated, the PCC between these two questionnaires was computed, see Table 8.10. The short PQ and Distraction factors are very little and negatively correlated. However, the short PQ and the other three factors (control, realism, and sensory) are positively correlated.

Table 8.11 Pearson correlation coefficient between integral quality, simulator sickness and presence for Test 12 (Singla et al. 2019)

Group	Integral quality short SSQ	Integral quality long SSQ (total score)	Integral quality long PQ (Realism)	Integral quality short PQ	Short SSQ long PQ (Realism)	Long SSQ (total score) short PQ
#1	−0.27	−0.34	0.78	0.57	−0.40	0.17
#2	0.01	−0.09	0.43	0.57	−0.26	0.13

From the rather small correlations, it can be concluded that PQ_S measures a certain sub-part of presence as assessed using PQ_L, and replacing the latter with the short version must be done with respective awareness of the limited results.

8.2.4 Relationship Between Integral Quality, Presence, and Simulator Sickness

The PCC was computed between integral quality, simulator sickness, and presence for each subject. Table 8.11 shows the PCC values between integral quality, simulator sickness, and presence averaged across all subjects. The results show that integral quality is negatively correlated with Short and Long SSQ for Group #1 which signifies that with an increase in the scores of simulator sickness, integral quality will decrease or the other way around. For Group #2, there is a poor correlation between integral quality and with Short and Long SSQ. With respect to the correlation between integral quality and Short and Long PQ, they are positively correlated irrespective of the group and type of questionnaire. In the presented test results, the relationship between simulator sickness and presence is not conclusive as the relation between PQ_L (Realism) and SSQ_S is negative and PQ_S and SSQ_L (Total Score) is positive, each with low values.

8.3 Conclusion

In this chapter, the suitability of single-scale versions of the simulator sickness and presence questionnaire was investigated, in comparison to the long versions. To this aim, integral quality, simulator sickness, and presence were assessed using different test protocols for two groups of subjects. Results show that the Short SSQ (SSQ_S) and Short PQ (PQ_S) can only be used to evaluate stronger effects such as the impact of content on simulator sickness and sense of presence, and replacing the long version with the short version must be done with respective awareness of the limited results.

Furthermore, the relationship between integral quality, simulator sickness, and presence was also studied. Results indicated that presence has a positive correlation with integral quality. In the presented test results, the relationship between simulator sickness and presence and integral quality and simulator sickness is not conclusive.

References

S.A. Balk, M.A. Bertola, V.W. Inman, Simulator sickness questionnaire: Twenty years later, in *Proceedings of the Seventh International Driving Symposium on Human Factors in Driver Assessment, Training, and Vehicle Design* (2013), pp. 257–263

K.S. Hale, K.M. Stanney, *Handbook of Virtual Environments: Design, Implementation, and Applications* (CRC Press, Boca Raton, 2014)

S. Ishihara, *The Series of Plates Designed as a test for Colour Deficiency* (1997). http://www.dfis.ubi.pt/~hgil/P.V.2/Ishihara/Ishihara.24.Plate.TEST.Book.pdf. 16 May 2018

ITU-R BT.500-13, *Recommendation ITU-R BT.500-13: Methodology for the Subjective Assessment of the Quality of Television Pictures* (2012)

ITU-T P.915, *Subjective Assessment Methods for 3D Video, P.915* (2016)

ITU-T Rec. P.910, *Recommendation ITU-T P.910: Subjective Video Quality Assessment Methods for Multimedia Applications* (2008)

R.S. Kennedy et al., Simulator sickness questionnaire: an enhanced method for quantifying simulator sickness. Int. J. Aviat. Psychol. **3**(3), 203–220 (1993)

A. Singla et al., Measuring and comparing QoE and simulator sickness of omnidirectional videos in different head mounted displays, in *9th International Conference on QoMEX* (2017), pp. 1–6. https://doi.org/10.1109/QoMEX.2017.7965658

A. Singla et al., Assessing media QoE, simulator sickness and presence for omnidirectional videos with different test protocols, in *2019 IEEE Conference on Virtual Reality and 3D User Interfaces (VR)* (2019), pp. 1163–1164. https://doi.org/10.1109/VR.2019.8798291

H.T.T. Tran et al., A subjective study on QoE of 360 video for VR communication, in *IEEE 19th International Workshop on MMSP* (2017), pp. 1–6

B.G. Witmer, M.J. Singer, Measuring presence in virtual environments: a presence questionnaire. Presence **7**(3), 225–240 (1998)

Chapter 9
Conclusion and Future Work

9.1 Summary of Results

- A new and high-quality set of seven 360° videos (with low-quality audio) are produced in this book. These sequences contain indoor, outdoor, sports, and nature scenes, covering a wide range of spatial properties and exploration behavior. The sequences are captured in 8K (7680 × 3840) resolution with each a duration of at least 30 seconds.
- The perceived quality of 360° videos was evaluated and compared using different test methods in several experiments. Based on the results, it is recommended to use the DSIS test method when a more fine-grained resolving power is sought. When somewhat less resolving power of the method is acceptable and the testing time is of relevance, then ACR should be used.
- Recommendations/Guidelines were derived for conducting tests for the systematic evaluation of video quality of 360° videos. Prior to this work, no such guidelines existed for the evaluation of 360° video quality. ITU-T P.919 (2020) was the first recommendation on the subjective test methods for 360° video—with contributions from the author.
- An extensive set of results on the impact of different technical factors on 360° video quality was presented based on a total of 12 tests.
- The influence of different factors such as video resolution, bitrate, session duration, and latency (in the case of tile-based streaming) on cybersickness was investigated. It was shown that, in a higher resolution video, users are less prone to simulator sickness, and users are more comfortable watching a higher bitrate video than in lower.
- The conceptual framework called *SiSiMo* (Raake et al. 2020) to develop models to predict simulator sickness scores was instantiated with a first model. It was demonstrated that the temporal evolution of discomfort can be estimated using linear regression over viewing-time index for 360° videos.

A. Singla, *Assessment of Visual Quality and Simulator Sickness for Omnidirectional Videos*, T-Labs Series in Telecommunication Services,
https://doi.org/10.1007/978-3-031-49988-3_9

- The SSQ was simplified to enable more efficient testing for 360° video. Based on the conducted analysis, it was shown that the resulting nine symptoms can be expected to capture all the relevant symptoms, just as the SSQ does, and hence can be used as an alternative to the SSQ for studies related to 360° videos.
- The Short SSQ (SSQ_S) and Short PQ (PQ_S) can only be used to evaluate stronger effects such as the impact of content on simulator sickness and sense of presence. To evaluate the impact of factors such as resolution, bitrate, frame rate, and impact of test duration, the long versions SSQ_L and PQ_L should be used.

9.2 Future Work

There are some limitations of the work presented in this dissertation that could be improved in several aspects. Some of these are provided below:

- As it can be seen from the state of the art (Chap. 2) and Sect. 3.3, the used 360° video sequences are without adequate high-quality audio. To investigate the audiovisual quality of 360° videos, there is a need for 360° scenes that contain high-quality spatial audio. Furthermore, most of the sequences were shot with a stationary camera. In order to more systematically investigate the symptoms of simulator sickness, there is a need that some of the used 360° video sequences to be shot with a moving camera.
- In Chap. 4, subjective tests were conducted using the ACR, M-ACR, and DSIS test methods in different contextual conditions. Most of the studies in the state of the art (Chap. 2) used the ACR test method to assess the visual quality of 360° videos. There are a few studies that use paired comparison and SAMVIQ. The work could be extended by conducting a subjective test under the same contextual conditions using all the above-mentioned test methods to find the most suitable and reliable test method, e.g., answering the question of how well different conditions can be distinguished, and how many subjects are required to do so (Brunnström and Barkowsky 2018).
- In this book, the impact of resolutions, bitrates, HMDs, and delay values (tile-based streaming) is investigated in Chap. 5 on the visual quality of 360° video. The impact of frame rate, motion interpolation algorithms, and encoders on the perceived visual quality is mentioned in Chap. 2 (State of the art). As, a 360° video can be represented in different projection formats such as ERP, Cubemap, Equal-area, Octahedron, Icosahedron, etc. Future research should be conducted to evaluate which of the projection schemes provides better subjective video quality.
- All of the studies used in this book revolve around 360° video, not VR. VR and 360° video differ in terms of Degrees of Freedom (DoF) and the way they are captured. Here, VR can provide movement to a user in 6 DoF (3 directional and 3 rotational). On the other hand, 360° video is captured from a real-life scene

using a 360° camera and can provide movement to a user in only 3 DoF (3 rotational). Additionally, in VR, a user can interact with the objects present in the scene. Therefore, it would be reasonable to explore, for the same scene, if VR can provide a better user experience than 360° video, as it provides more degrees of freedom, and a user can interact with the objects as well. It should be noted that user experience in VR may be affected by the realness of the visual representation of the computer-generated world.

- The SSQ was mainly used for the assessment of simulator sickness. The SSQ was adapted for VR contents by Kim et al. (2018) and Stone (2017), but not for 360° contents. In Chap. 7, a simplified version of the SSQ is proposed that is relevant for 360° video related studies. Please note that it is not argued that the SSQ cannot be used in the case of 360° videos. Future work could be extended by conducting concrete tests where the proposed 9-attribute questionnaire will need to be verified with the reduced set of questions and the SSQ.
- In Chap. 8, results show that the single-scale questions do not have a very high correlation with the longer version of questions. To improve the correlation between those two versions of questions, better instruction before the start of the test could be provided to subjects related to the usage of the single-scale questions. This may help to improve the correlation between single-scale and longer-version questions for simulator sickness and presence.
- In Chap. 8, only one single scale for the evaluation of presence and simulator sickness was compared with the long versions. In Gutierrez et al. (2022), two single-scale questions for the evaluation of simulator sickness were compared with the long SSQ. However, in the state of the art (Chap. 2), there are a few more single-scale questions that could be compared with the long SSQ. Thus, future work could be extended by comparing different single-scale questions with the long version of the SSQ for the assessment of simulator sickness, to find the best possible single-scale question that can evaluate the impact of resolution, bitrate, and test duration. Furthermore, dedicated tests need to be conducted to find the inter-relation between simulator sickness and presence and integral quality and simulator sickness.

References

K. Brunnström, M. Barkowsky. Statistical quality of experience analysis: on planning the sample size and statistical significance testing. J. Electron. Imag. **27**(5), 053013-1–11 (2018)

J. Gutierrez et al., Subjective evaluation of visual quality and simulator sickness of short 360° videos: ITU-T Rec. P.919. IEEE Trans. Multimedia **24**, 3087–3100 (2022). https://doi.org/10.1109/TMM.2021.3093717

ITU-T P.919, *Subjective Test Methodologies for 360° Video on Head-Mounted Displays*. Recommendation P.919. Oct. 2020

H.K. Kim et al., Virtual reality sickness questionnaire (VRSQ): motion sickness measurement index in a virtual reality environment. Appl. Ergonom. **69**, 66–73 (2018). ISSN: 0003-6870. https://doi.org/10.1016/j.apergo.2017.12.016. http://www.sciencedirect.com/science/article/pii/S000368701730282X

A. Raake et al., SiSiMo: Towards simulator sickness modeling for 360- degree videos viewed with an HMD, in *2020 IEEE Conference on Virtual Reality and 3D User Interfaces Abstracts and Workshops (VRW)*. (2020), pp. 582–583

W.B. Stone III, *Psychometric Evaluation of the Simulator Sickness Questionnaire as a Measure of Cybersickness*. Doctoral dissertation, Iowa State University, 2017

Appendix A
List of Questionnaires

A.1 Video Quality Questions

Test 1 and 12
Integral quality of VR presentation

Excellent	Good	Fair	Poor	Bad
5	4	3	2	1

Test 2, 4, and 5
How would you rate the quality of this video?

Excellent	Good	Fair	Poor	Bad
5	4	3	2	1

Test 3 and 6
Please rate the degradations of this video w.r.t. the reference video on a five-point scale.

Imperceptible	Perceptible but not annoying	Slightly annoy-ing	Annoying	Very annoy-ing
5	4	3	2	1

A. Singla, *Assessment of Visual Quality and Simulator Sickness for Omnidirectional Videos*, T-Labs Series in Telecommunication Services,
https://doi.org/10.1007/978-3-031-49988-3

Test 8 and 9
How would you rate the quality of the video?

0	1	2	3	4	5	6	7	8	9	10
	Bad		Poor		Fair		Good		Excellent	

Test 10
How would you rate the audiovisual quality of the video?

Excellent	Good	Fair	Poor	Bad
5	4	3	2	1

Test 11
How would you rate the quality of this 360° video?

Excellent	Good	Fair	Poor	Bad
5	4	3	2	1

A.2 Short Simulator Sickness Questions

Test 8, 9, and 12
How is the level of dizziness or nausea during VR viewing experiment on the scale from 1 to 5?

Absolutely not dizzy	Not dizzy	Slightly dizzy	Dizzy	Very dizzy
1	2	3	4	5

Test 10
How would you rate your comfort level while watching the VR session?

Very comfortable	Comfortable	Mildly uncomfortable	Uncomfortable	Extremely uncomfortable
5	4	3	2	1

Test 11
How is the level of dizziness or nausea on the scale from 1 to 5?

Absolutely not dizzy	Not dizzy	Slightly dizzy	Dizzy	Very dizzy
1	2	3	4	5

Are you feeling any sickness or discomfort now? Check the appropriate sentence

No problem	No perceptible effect, natural feeling
Light effects	Slight discomfort, but no sickness
Uncomfortable	Moderate discomfort, but tolerable for a while
Unpleasant	Strong discomfort or sickness, but can continue the test
Unbearable	Strong discomfort or sickness, and want to stop test

A.3 Simulator Sickness Questionnaire

1. General discomfort

None	Slight	Moderate	Severe

9. Difficulty concentrating

None	Slight	Moderate	Severe

2. Fatigue

None	Slight	Moderate	Severe

10. Fullness of the head

None	Slight	Moderate	Severe

3. Headache

None	Slight	Moderate	Severe

11. Blurred vision

None	Slight	Moderate	Severe

4. Eye strain

None	Slight	Moderate	Severe

12. Dizziness with eye open

None	Slight	Moderate	Severe

5. Difficulty focusing

None	Slight	Moderate	Severe

13. Dizziness with eye closed

None	Slight	Moderate	Severe

6. Salivation increasing

None	Slight	Moderate	Severe

14. Vertigo

None	Slight	Moderate	Severe

7. Sweating

None	Slight	Moderate	Severe

15. Stomach awareness

None	Slight	Moderate	Severe

8. Nausea

None	Slight	Moderate	Severe

16. Burping

None	Slight	Moderate	Severe

A.4 Presence Questionnaire

1. How responsive was the environment to actions that you initiated (or performed)?

Not responsive			Moderately responsive			Completely responsive

2. How natural did your interactions with the environment seem?

Extremely artificial			Borderline			Completely natural

3. How completely were all of your senses engaged?

Not engaged			Mildly engaged			Completely engaged

4. How much did the visual aspects of the environment involve you?

Not at all			Somewhat			Completely

5. How much did the auditory aspects of the environment involve you?

Not at all			Somewhat			Completely

6. How aware were you of events occurring in the real world around you?

Not aware at all			Mildly aware			Very aware

7. How aware were you of your display?

Not aware at all			Mildly aware			Very aware

8. How compelling was your sense of objects moving through space?

Not at all			Moderately compelling			Very compelling

9. How inconsistent or disconnected was the information coming from your various senses?

Not at all inconsistent			Somewhat inconsistent			Very inconsistent

10. How much did your experiences in the virtual environment seem consistent with your real world experiences?

Not consistent			Moderately consistent			Very consistent

11. How completely were you able to actively survey or search the environment using vision?

Not at all			Somewhat			Completely

12. How well could you identify sounds?

Not at all			Somewhat			Completely

13. How well could you localize sounds?

Not at all			Somewhat			Completely

14. How compelling was your sense of moving around inside the virtual environment?

Not compelling			Moderately compelling			Very compelling

15. How closely were you able to examine objects?

Not at all			Pretty closely			Very closely

16. How involved were you in the virtual environment experience?

Not involved			Mildly involved			Completely engrossed

17. How much delay did you experience between your actions and expected outcomes?

No delays			Moderate delays			Long delays

18. How quickly did you adjust to the virtual environment experience?

Not at all			Slowly			Less than 30 seconds

19. How proficient in moving and interacting with the virtual environment did you feel at the end of the experience?

Not proficient			Reasonably proficient			Very proficient

20. How much did the visual display quality interfere or distract you from performing assigned tasks or required activities?

Not at all			Interfered somewhat			Prevented performance

21. Were you involved in the experimental task to the extent that you lost track of time?

Not at all			Somewhat			Completely

22. To what degree did you feel confused or disoriented at the beginning of breaks or at the end of the experimental session?

Not at all			Mildly disoriented			Very disoriented

A.5 Short Presence Question (Test 12)

How is your assessment about the sense of presence in the VR environment on a scale from 1 to 5? (1 means that you have absolutely no sense of presence, and 5 means that you have a true sense of presence as in a real environment.)

High				Low
5	4	3	2	1

Index

A. Singla, *Assessment of Visual Quality and Simulator Sickness for Omnidirectional Videos*, T-Labs Series in Telecommunication Services,
https://doi.org/10.1007/978-3-031-49988-3

The manufacturer's authorised representative in the EU is Springer Nature Customer Service Centre GmbH, Europaplatz 3, 69115 Heidelberg, Germany. If you have any concerns regarding our products, please contact ProductSafety@springernature.com

Printed and bound by CPI Group (UK) Ltd, Croydon, CR0 4YY
30/09/2026
02225309-0011